Lecture Notes in Computer Scien

Edited by G. Goos, J. Hartmanis and J. van Lee

Springer
Berlin
Heidelberg
New York
Barcelona
Budapest
Hong Kong
London
Milan
Paris
Singapore
Tokyo

Edmund Burke   Michael Carter (Eds.)

Practice and Theory of Automated Timetabling II

Second International Conference, PATAT'97
Toronto, Canada, August 20-22, 1997
Selected Papers

 Springer

Series Editors

Gerhard Goos, Karlsruhe University, Germany
Juris Hartmanis, Cornell University, NY, USA
Jan van Leeuwen, Utrecht University, The Netherlands

Volume Editors

Edmund Burke
University of Nottingham, Department of Computer Science
University Park, Nottingham NG7 2RD, UK
E-mail: ekb@cs.nott.ac.uk

Michael Carter
University of Toronto
Department of Mechanical and Industrial Engineering
5 King's College Road, Toronto, Ontario, Canada M5S 3G8
E-mail: carter@mie.utoronto.ca

Cataloging-in-Publication data applied for

Die Deutsche Bibliothek - CIP-Einheitsaufnahme

Practice and theory of automated timetabling II : second
international conference ; selected papers / PATAT '97, Toronto,
Canada, August 20 - 22, 1997 / Edmund Burke ; Michael Carter
(ed.). - Berlin ; Heidelberg ; New York ; Barcelona ; Budapest ; Hong
Kong ; London ; Milan ; Paris ; Singapore ; Tokyo : Springer, 1998
  (Lecture notes in computer science ; 1408)
  ISBN 3-540-64979-4

CR Subject Classification (1991): I.2.2, I.2.8, H.4.1-2, J.1, K.3, F.2, G.2

ISSN 0302-9743
ISBN 3-540-64979-4 Springer-Verlag Berlin Heidelberg New York

© Springer-Verlag Berlin Heidelberg 1998
Printed in Germany

Typesetting: Camera-ready by author
SPIN 10637176   06/3142 – 5 4 3 2 1 0   Printed on acid-free paper

Preface

This volume contains the selected papers from the second international conference on the Practice And Theory of Automated Timetabling (PATAT'97) held in Toronto, August 20-22, 1997.

The Concise Oxford Dictionary (seventh edition) defines a "time table" as a "scheme of school work, etc., table showing times of public transportation". Most dictionaries do not include "timetabling" as a word. "Timetabling" can be defined as the process of assigning specific times to a set of activities. It occurs in a wide variety of organisations, institutions and everyday situations. Everybody has come into contact with a timetable at one time or another in their lives. Common examples occur in universities and schools (determine what time each exam will be held or what time each course will take place); in transportation (determine what time a train/bus/aeroplane will depart/arrive for each station/airport) and in sports (determine the time and date of each match/event to be played). Timetabling is also required for conferences (determine the times for each session), for interview planning (determine which dates each party will be available for interviews) and for employee management (designing each employee's work timetable).

The term timetabling and the term scheduling are sometimes used interchangably. People will talk about timetabling a set of exams or scheduling a set of exams. Anthony Wren* surveyed the use of the terms scheduling, timetabling and rostering, and found a great deal of variety. He decided to define timetabling as a special case of scheduling:

"Timetabling is the allocation, subject to constraints, of given resources to objects being placed in space-time, in such a way as to satisfy as nearly as possible a set of desirable objectives."

For example, examinations and courses must be scheduled so that no person should be expected to be in two places at once! In addition, there may be space constraints on the size of rooms available at any time, and special facilities required. There may also be a number of features that are desirable but not absolutely critical in all cases such as attempting to reduce the number of times a student has to take two exams consecutively (without a break). Indeed the problem varies from situation to situation. Some educational institutions will place more weight on constraints like these than other institutions do.

People have been constructing timetables for hundreds of years, and yet we know surprisingly little about efficient techniques for performing this class of tasks. Even today, the majority of timetabling functions are carried out manually. In some cases, it is done by someone who, through years of experience, has developed an intuitive sense of how the pieces should fit together. However,

---

* See A. Wren, "Scheduling Timetabling and Rostering - A special Relationship", in Burke and Ross (eds.), "The Practice and Theory of Automated Timetabling", Lecture Notes in Computer Science, Vol. 1153, 1996, pages 46-76.

more often than not, timetabling is left to relative amateurs who spend days and weeks trying to devise a reasonable solution. Yet, research papers on automated timetabling have consistently appeared over the last forty years or so and in many situations, computerised methods have been used to assist in the problem solution. However, it is clear that there is still much work to be done on the automation of this important class of problems. Indeed, in the last few years, the level of research interest in this fascinating area has increased dramatically. The establishment of the series of international conferences on the Practice and Theory of Automated Timetabling (PATAT) and the great success of the first two conferences in the series have reflected this increase of interest in the area. Another indicator is the establishment of the EURO (Association of European Operational Research Societies) Working group on Automated TimeTabling (WATT). See http://www.cs.rdg.ac.uk/cs/research/pedal/watt/index.html for further details.

This volume has been organised into sections based on the primary solution method employed in the papers (Tabu search and simulated annealing, population based methods, constraint based methods and graph theory). Of course, there is some overlap in the categorisations and some papers could happily have fitted into two (or more) sections. We also include a section on surveys, and a section on practical issues. Every effort has been made to ensure that among the papers in the conference (and so in this volume of selected papers), there will be something of interest to both researchers and practitioners. We have, therefore, in addition to research papers, included papers on practical issues that describe some of the more subjective considerations in timetabling from the viewpoint of administrators and practitioners. The main motivation behind this is, not only that the papers will be of interest to practitioners, but also that they should be of interest to researchers in the area and should help them to focus their research on some of the issues that administrators raise.

The Conference Series

The conference in Toronto was the second in a series of international conferences on the Practice And Theory of Automated Timetabling (PATAT). This volume contains the selected papers from that conference.

The first conference was held in Edinburgh in August/September 1995. Selected papers from this also appeared in the Springer Lecture Notes in Computer Science series. The full reference is:

Edmund Burke and Peter Ross (editors), The Practice and Theory of Automated Timetabling: Selected Papers from the 1st International Conference on the Practice and Theory of Automated Timetabling, Edinburgh August/September 1995, Lecture Notes in Computer Science, Vol. 1153, Springer 1996.

The third conference in the series will be held in Konstanz, Germany, in August 2000. Future conferences will be held every three years. For further information about the conference series, contact the steering committee (see below) or see http://www.asap.cs.nott.ac.uk/ASAP/ttg/patat-index.html

Acknowledgements

The conference in Toronto was a great success. It built on the considerable success of the first conference in the series. Many people worked very hard to make the conference run smoothly and efficiently. The organising committee (members are listed below) worked extremely hard to make it the valuable, enjoyable and interesting event that it turned out to be. We would like to extend our sincere thanks to them. Particular thanks go to Margaret Thompsett. She looked after registration, notification, correspondence, meals and residence, and virtually ran the local arrangements single-handedly. Under her capable guidance, everything went smoothly.

The papers and abstracts were fully refereed for the conference itself. All the papers that were submitted to this "Selected papers" volume had to undergo a second round of very careful and rigorous refereeing. A lot of work (throughout both rounds of refereeing) has gone into the process of selecting those papers which now appear in this book. We are very grateful to the members of the programme committee (listed below) who helped to referee the papers during these two rounds. Thanks must also go to the staff of Springer for their support and encouragement. As editor of the Lecture Notes in Computer Science series, Jan van Leeuwen was particularly helpful throughout this project, as he was with the previous volume. His comments and advice were, once again, invaluable in bringing the book to publication. We are also particularly grateful to Farida Alibhai for the secretarial work she carried out during the first round of refereeing and to Carol Jones for her work during the (early part of the) second round of refereeing. Special thanks also go to Alison Payne for all the secretarial support she has given us towards the end of the second round of refereeing and during the preparation of this volume.

The main contributors to the success of the conference were (once again), without doubt, the delegates, the vendors who presented their software and the authors of submitted papers and presentations. Our thanks go to them for the enthusiasm and support they have given to us and to this event and that (hopefully) they will continue to give to future events. Finally, we would like to thank the steering committee (listed below) for their continuing work in bringing us this and future PATAT conferences. We apologise for any omissions that have been inadvertently made. So many people have helped with this conference and with the series of conferences that it is difficult to remember them all.

April 1998   Edmund Burke
                                                              Mike Carter

Second International Conference on the Practice and Theory of Automated Timetabling (PATAT '97) Programme Committee

Second International Conference on the Practice and Theory of Automated Timetabling (PATAT '97) Organising Committee

Department of Mechanical and Industrial Engineering
University of Toronto, Canada

Michael Carter	Chair
Cathie Kessler	Registration desk
Maurine Kwok	Finance
Sau Yan Lee	Technical support
Ella Lund-Tompsen	Registration desk
Wendy Smith	Secretarial support
Margaret Tompsett	Everything and anything else

The International Series of Conferences on the Practice And Theory of Automated Timetabling (PATAT) Steering Committee

Edmund Burke (Chair)	University of Nottingham, UK
Ben Paechter (Treasurer)	Napier University, UK
Victor Bardadym	International Renaissance Foundation, Ukraine
Michael Carter	University of Toronto, Canada
Dave Corne	University of Reading, UK
Wilhelm Erben	Fachhochschule Konstanz – University of Applied Sciences, Germany
Jeff Kingston	University of Sydney, Australia
Gilbert Laporte	Ecole des Hautes Etudes Commerciales, Montreal, Canada
Peter Ross	University of Edinburgh, UK
George White	University of Ottawa, Canada
Dominique de Werra	EPFL, Switzerland
Anthony Wren	University of Leeds, UK

Contents

Constraint Based Methods

Graph Theory

Practical Issues

Other Timetabling Presentations

Author Index

Surveys

Recent Developments in Practical Course Timetabling

Michael W. Carter[1] and Gilbert Laporte[2]

[1]Department of Mechanical and Industrial Engineering
University of Toronto
5 King's College Road
Toronto, Ontario M5S 3G8
carter@mie.utoronto.ca

[2]GERAD
École des Hautes Études Commerciales de Montréal
3000 chemin de la Côte-Sainte-Catherine
Montréal, Canada H3T 1V6
gilbert@crt.UMontreal.ca

Abstract. Course timetabling is a multi-dimensional NP-Complete problem that has generated hundreds of papers and thousands of students have attempted to solve it for their own school. In this paper, we describe the major components of the course timetabling problem. We discuss some of the primary types of algorithms that have been applied to these problems. We provide a series of tables listing papers in refereed journals that have either implemented a solution or tested their algorithm on real data. We made no attempt to provide a qualitative comparison. We restricted our presentation to a description of the types of technique used and the size of problem solved We have not included commercial software vendors

1   Introduction

Course timetabling is an important problem encountered in virtually every high school, college and university throughout the world. Over the past thirty years or so, several algorithms have been proposed for the solution of this problem. Our purpose is to report on some recent developments and trends in this area with an emphasis in work that has actually been implemented. More specifically, as in Carter and Laporte [10], we will concentrate on articles published since 1980 and on those that report on real implementations or on tests performed on real data. There are a number of timetabling systems in practice and in the literature that use the computer as a decision support tool for a primarily manual process. We have restricted our survey to include only those papers where the computer is actually used to attempt to solve the timetabling problem automatically.

This survey is organized as follows. In Section 2, we describe the various components of the course timetabling problem. This is followed in Section 3 by a description of the main algorithms used in this field. In Section 4, we summarize in table form the main algorithms that have either been implemented, or tested on real data. The conclusions follow on Section 5.

2   Problem Components

Course timetabling can be viewed as a multi-dimensional assignment problem in which students, teachers (or faculty members) are assigned to courses, course sections or classes; "events" (individual meetings between students and teachers) are assigned to classrooms and times. Problem definition and terminology varies from one institution to the next but the following concepts are common.

a)   A *course* corresponds to a subject taught one or more times a week during part of a year. The same course may be taught in one or multiple sections, meaning that it can be repeated by different teachers during the week.

b)   A *class* is a group of students taking an identical set of courses and typically remaining together throughout the week.

c)   A *program* consists of a set of required courses and a set of elective courses to be taken by students wishing to obtain a given degree.

Some of these concepts do not apply to some institutions. In particular, high schools often work with predefined classes and have few programs so that the main problem in this context is to determine teachers for given subjects and classes, and to construct appropriate teaching schedules. The time frame is typically limited and teachers have heavy teaching loads. The scheduling problem is therefore very tight. In universities, individual students have more choice and faculty members may only teach a few hours a week. The main difficulty in this case is to ensure that students can feasibly select courses required by their program. We summarize in Table 1, the major differences between the high school and the university scheduling problems. Clearly, there are high schools with a variety of electives, and universities with structured problems; but, we present these classical definitions for extreme points on the continuum.

The course scheduling problem is normally decomposed into five different subproblems, not all of whom may be relevant to a particular situation:

Table 1. Characteristics of 'High School' vs 'University' Course Timetabling

Characteristic	High School	University
Scheduling	- by classes	- by student
Choice	- few choices - highly structured programs	- many electives - loosely structured programs
Teacher Availability	- tight (heavy teaching load)	- flexible (light teaching load)
Rooms	- few rooms - same size - central location	- many rooms - variety of sizes - decentralized
Student Load	- very tight (busy all day)	- fairly loose - days and evenings
Criteria	- no conflicts	- minimum conflicts

2.1 Course Timetabling

Assign courses or course sections to time periods. Several side constraints may be present such as respecting program structures as well as room and teacher availabilities.

In some instances, courses or course sections must be spread in particular ways throughout the week. For example, some institutions require that all sections of a same course be scheduled at the same times. A course may be broken into two or three periods and these must be scheduled on different days, etc. Some schedules include lunch periods, time to travel between different buildings, and constraints may be imposed on the number of teaching hours a student or teacher can have in a given day.

We also distinguish course timetabling problems as *Master Timetables* or *Demand Driven* systems. The primary difference between these systems lies in the sequence in which the various subproblems are solved.

Master Timetabling

1. Determine number of sections
2. Assign times (and rooms) to sections
3. Students choose courses from published timetables
4. Students assigned to sections (student scheduling)
5. Drop/add "negotiation" phase

Demand-Driven

1. Students select courses from list
2. Determine number of sections
3. Assign teachers to sections
4. Assign times (and rooms) to sections
5. Student scheduling (assign students to sections)
6. Drop/add

2.2 Class-Teacher Timetabling

This problem arises mostly in high schools. Here, the scheduling unit is a class, not a student. It is normally assumed that the assignment of teachers to courses and classes has already been made. The problem is then to schedule class-teacher meetings without creating conflicts and while satisfying some side constraints on the spread and sequencing of courses. Feasibility is often a major difficulty there and it is common to reconsider previously made class-teacher assignments in order to reach a feasible solution. It is interesting to observe, however, that the class-teacher timetabling problem, without side constraints, can be solved in polynomial time by means of a network flow algorithm (de Werra [20]).

2.3 Student Scheduling

This problem occurs when courses are taught in multiple sections. Once students have selected their courses, they must be assigned to sections. The aim is to provide good quality (conflict free) schedules for students while balancing section sizes and respecting room capacities.

2.4 Teacher Assignment

The problem here is to assign teachers to courses while maximizing a preference function. Again, this problem without side constraints can be solved as a network flow problem (Dryer and Mulvey [22]).

2.5 Classroom Assignment

Events must be assigned to specific rooms to satisfy size, location and facility preferences and restrictions. This is normally performed after the individual events have been assigned to times. Ideally, times and rooms should be assigned simultaneously, but many schools separate the two processes for simplicity.

3 Algorithms

The algorithms used to solve the various components of the course timetabling problem can broadly be classified into global algorithms, constructive heuristics and improvement heuristics. In addition, several methods include user interaction with a computer system.

3.1 Global Algorithms

Small size problems can sometimes be solved directly by means of a standard integer linear programming (ILP) package. Side constraints can often be incorporated into the ILP formulation. Several decomposition techniques can be used for problems that are too large for optimizing. Examples are provided by Glassey and Mizrach [26], and Carter [7]. In the Gosselin and Truchon [27] method for the classroom assignment problem, an ILP is solved repeatedly with different cost coefficients to account for various preference values. In most ILP's, 0-1 variables correspond to assignment decisions. An interesting exception is the McClure and Wells [37] method for the teacher assignment problem in which each variable represents a full teacher schedule and the problem is formulated as a set partitioning problem with side constraints. The number of variables is restricted in order to reduce problem size.

In a similar vein, network flow formulations and algorithms are sometimes used to solve some simple assignment problems without side constraints. Examples are provided in the articles of Osterman and de Werra [42] and Chahal and de Werra [11] for class-teacher assignment, and of Dyer and Mulrey [22] and Dinkel, Mote and Venkataramanan [21] for teacher assignment.

Preferences can be handled by goal programming, as in the work of Miyaji, Ohno and Mine [38] for student scheduling and of Schniderjans and Kim [54] for teacher assignment. Alternatively, an assignment problem can be solved repeatedly with different weights. Graves, Schrage and Sankaran [28] describe an auction based interactive system for student scheduling that uses this idea.

3.2 Constructive Heuristics

The simplest heuristic for constrained assignment problems probably consists of making sequential assignments while preserving feasibility, until this is no longer possible. Then, backtracking may be used to undo some of the previous decisions and the assignment process is reapplied. This type of method may not, however, always converge toward a feasible solution and care must be taken to avoid cycling. This is essentially the method used by Carter, Laporte and Lee [9] for examination timetabling. In the field of course timetabling, it has been applied by Papoulias [44], de Gans [19] and by Cooper and Kingston [15] for class-teacher timetabling, by Sabin and Winter [51] for student scheduling, and by Selim [55] for teacher assignment.

Incomplete branch and bound has been used by Cooper and Kingston [15] for class-teacher timetabling, and by Laporte and Desroches [36] for student scheduling.

More sophisticated enumerative algorithms are based on constraint logic programming (CLP) or similar methods. Here the problems are modeled using variables with finite domains. As variables are fixed, the domains of the free variables are restricted. This technique has been gaining popularity in recent years, as witnessed by the work of Azevedo and Barahona [5], Cheng, Kang, Leung and White [13], Ram and Scogings [46], Guéret, Jusien, Boizumault and Prins [29], Lajos [35], and Henz and Wurtz [30] for course timetabling, and of Fahrion and Dollansky [25] for teacher assignment.

3.3 Improvement Heuristics

Once a feasible solution has been obtained, it can be improved by means of a standard exchange scheme as in the work of Aubin and Ferland [4] and Sampson, Freeland and Weiss [52] for course timetabling. In classical improvement procedures, the search only proceeds from a solution to a better solution. However, with modern search methods such as simulated annealing (SA) and tabu search (TS), the search may move towards a worse solution. Both methods consider at iteration t a solution x_t and its neighbourhood $N(x_t)$. A neighbour of x_t is any solution derived from x_t by applying a given operation (such as switching two elements, etc.) In SA, a candidate solution x is randomly selected in $N(x_t)$. If x improves upon x_t , then x_{t+1} , is set equal to x. Otherwise, x_{t+1} is set equal to x with probability P_t and to x_t with probability $(1-P_t)$, where P_t is a decreasing function of t. In TS, some solutions are declared tabu during the search process. At iteration t, x_{t+1} is set equal to the best non-tabu solution of $N(x_t)$ and x_t is then declared tabu for a number of iterations.

Evolutionary or genetic algorithms (GA) constitute another popular search strategy. Here a population of solutions is maintained. At every iteration, two parent solutions are selected to produce two offspring and two elements of the population are discarded. The idea is to produce better and better offspring by selecting some of the best parents from the population. Several variants of these very basic SA, TS and GA algorithms have been produced over the years for a wide variety of combinational optimization problems. Interested readers can consult the books of Reeves [48], of Aarts and Lenstra [3] and the bibliography compiled by Osman and Laporte [41].

In the field of course timetabling, SA has been applied by Abramson [1] for class-teacher timetabling and by Davis and Ritter [18] for student scheduling. TS algorithms have been implemented by Hertz [31] for course timetabling, by Schaerf [53] and by Costa [17] class-teacher timetabling. A large number of GA algorithms have proposed. These include the work of Corne and Ross [16], Paechter, Cumming and Luchian [43], Rich [49] and Erben and Keppler [24] for course timetabling, and of Colorni, Dorigo and Maniezzo [14] for class-teacher timetabling.

3.4 Interactive Systems

In addition to the algorithmic strategies just described, several researchers propose interactive procedures to produce families of solutions under different constraints or preference weights. Chahal and de Werra [11] propose such a system for class-teacher timetabling and Graves, Schrage and Sankaran [28] use an interactive auctioning algorithm for student scheduling.

4 Summary Tables

We have summarized our findings in five tables. Table 2 lists papers related to Course Timetabling, Table 3 describes Class-teacher Timetabling, Table 4, 5 and 6 list the results for Student Scheduling, Teacher Assignment and Classroom Assignment respectively. For each problem, we list the references chronologically. In the second column, we provide the name of the institution and the size (where available) in terms of number of students, courses, faculty, sections, rooms and classes. The third column provides a brief description of the problem that the authors solved in their version. Column 4 describes the algorithm that was used, and column 5 states whether or not the method was implemented or simply tested on real data.

Table 2. Course Timetabling

Reference	Institution (students, courses, faculty, sections, rooms)	Problem Description	Algorithm	Results
Aubin & Ferland [4]	A large Montreal High School and two University departments	- course timetabling and sectioning - minimize student conflicts subject to lecturer and room availability - quadratic assignment	- exchange heuristic (hill climbing)	Implemented
Hertz [31]	University of Geneva Faculty of Ecomonics (1729,288,143,-,67)	- minimum faculty and student conflicts	- Tabu Search (TATI)	Tested on real data
Kang, van Schoenberg & White [34] Cheng, Kang, Leung & White [13]	University of Ottawa (-,2147,-,-,167)	- students grouped in programs with required and elective courses - assign courses to times, rooms, teachers	- constraint logic programming	Tested on real data
Azevedo & Barahona [5]	New University of Lisbon Computer Science Department (-,-,-,10)	- courses to times subject to room and teacher constraints	- expert system - constraint logic programming	Tested on real data
Ross, Corne & Fang [50] Corne & Ross [16]	University of Edinburgh, Department of Artificial Intelligence (-, 82, -,-,-)	- minimize lecturer conflict - minimize expected student conflict (lecturers in same theme) - room assignments	-evolutionary algorithm	Tested on real data
Paechter, Cumming & Luchian [43] (Rankin [47])	Napier University Computer Science (-,238,52,-,7)	- assign classes to timeslot subject to room, teacher and student group availability	-evolutionary algorithm (genetic)	Implemented

Table 2. Course Timetabling *Continued*

Reference	Institution (students, courses, faculty, sections, rooms)	Problem Description	Algorithm	Results
Sampson, Freeland & Weiss [52]	University of Virginia, Darden Graduate School of Business (230,64,-,89,-)	- determine times for course sections - assign to term - minimize student conflicts - satisfy teacher preferences - sectioning	local search: assign students one-at-a-time	Implemented
Erben & Keppler [24]	Unspecified University	- schedule courses and assign them to rooms subject to several side constraints	- genetic algorithm	Tested on data from a real context
Guéret, Jussien, Boizumault & Prins [29]	Institute of Applied Mathematics, France (160,91,42,-,8)	- schedule courses and assign them to rooms, subject to side constraints on rooms, teachers availabilities, student schedules, course interactions, etc.	- constraint logic programming	Tested in the 1993-1994 data
Henz & Würtz [30]	Catholic College for Social Work, Saarbrücken, Germany (-,91,34,-,7)	- schedule courses and assign them to rooms subject to thirteen classes of side constraints	- constraint logic programming	Tested on real data
Lajos [35]	University of Leeds (-,1000,-,2500,-)	- schedule courses and assign them to rooms, subject to seven classes of side constraints	- constraint logic programming	Leeds is looking at a follow-up project
Ram & Scogings [46]	University of Natal, South Africa (-,450,-,-,80)	- schedule courses and assign them to rooms, subject to side constraints on rooms	- multiple constraint reasoning	3 variations tested on real data
Rich [49]	A University (-,101,56,-,55)	- schedule courses and assign them to rooms	- genetic algorithm	Tested on real data
Elmohamed, Coddington & Fox [23]	Syracuse University (13653, 3839, -, 3839, 509)	- schedule courses and assign them to rooms subject to several side constraints	- simulated annealing	Tested on real data
White & Zhang [56]	Some U. of Ottawa depts (-, 262, -, -, 46)	- schedule courses and assign them to rooms subject to several side constraints	- Constraint logic and tabu search	Tested on real data

Table 3. Class-Teacher Timetabling

Reference	Institution (classes, courses, faculty, rooms)	Problem Description	Algorithm	Results
Papoulias [44]	Five British Schools (-,-,-,-)	- schedule class-teacher meetings with side constraints	- assignment technique with backtracking	Tested on real data
de Gams [19]	21 H.S. in the Netherlands (-,-,-,-)	- schedule class-teacher meetings with side constraints	- random assignments with some backtracking	Tested on real data
Ostermann & de Werra [42]	H.S.'s in Canton de Vand, Switzerland (40,1350,95,-)	- schedule class-teacher meetings	-network flow algorithm to create a partial schedule; completed by hand.	Implemented in several schools
Chahal & de Werra [11]	An Adult Training School in Geneva (-,-,-,-)	- schedule class-teacher meetings	- network flow to create partial schedule plus facilities to produce alternative schedules under different constraints	Implemented
Abramson [1]	An Australian H.S. (15 to 101, 100 to 757,15 to 37, 15 to 24)	- schedule class-teacher meetings with side constraints	- simulated annealing	Tested on four data sets
Colorni, Dorigo & Maniezzo [14]	A H.S. in Milan	- schedule class-teacher meetings with side constraints	- genetic algorithm	Tested on one data set. No details
Cooper & Kingston [15]	A H.S. in Sidney, Australia	- schedule class-teacher meetings with side constraints	- a variety of techniques are described including incomplete branch and bound, and assignment techniques with backtracking	Tested on real data set. No details
Costa [17]	A H.S. in Porrentrug, Switz. (32,780,65,12 types)	- schedule class-teacher meetings with side constraints	- tabu search	Tested on real data
Schaerf [53]	Guinto O. Flacco H.S., Potenza, Italy (38,27 to 29 lectures per class, -,-)	- schedule class-teacher meetings with some side constraints	- tabu search	Implemented
Chan, Lau & Sheung [12]	Several primary &H.S.'s (-, 1000, -, -)	- schedule class, teacher and rooms	- constraint satisfaction	Implemented
Nepal, Melville, Ally [39]	ML Sultan Technicon Faculty of Commerce (35,35,31,21)	- schedule class, teacher and rooms	- brute force heuristics	Tested on real data

Table 4. Student Scheduling

Reference	Institution (students, courses, sections)	Problem Description	Algorithm	Results
Laporte & Desroches [36]	École Polytechnique de Montréal (EPM) (2799,326,518)	Timetable is given. Assign students to course sections. Maximize student satisfaction. Balance section enrollments. Respect room capacities.	Incomplete branch and bound followed by two improvement phases.	Used by EPM starting in 1984
Sabin & Winter [51]	Memorial University, Newfoundland (-,-,-)	Timetable is given. Assign students to course sections. Maximize student satisfaction. Balance sections enrollments. Respect room capacities.	Construct potential student timetables by giving weights to course sections. Schedule students in a greedy fashion starting with those with the most awkward timetables. Feasibility always maintained in practice.	Implemented at Memorial University
Davis & Ritter [18]	Harvard University (118,-,-)	Assign students to sections while respecting preferences.	Simulated annealing.	Tested on real data
Miyaji, Ohno & Mine [38]	Okayama University, Japan (40 to 83, 13,-)	Assign students to sections.	Goal programming.	Tested on 41 real examples
Graves, Schrage & Sankaran [28]	University of Chicago Graduate School of Business (GSB) (>2000,-,475)	Allocate course sections to students while respecting student preferences. Respect section capacities. Allows for preference-based changes of registration during the term of registration.	Interactive system. Students use bidding points to bid on desired course schedules. Drop, add and swap operations are possible.	Implemented at the GSB since 1981

Table 5: Teacher Assignment

Reference	Institution (courses, faculty)	Problem Description	Algorithm	Results
Dyer & Mulvey [22]	UCLA, Graduate School of Management (-,-)	Assign faculty members to courses. Respect availabilities and preferences.	Network optimization model and algorithm.	Implemented since 1974
Bloomfield & McSharry [6]	Oregon State University School of Business (-)	Assign teachers to classes or section. Assign days, times, rooms to sections. Criteria: teacher preferences and seniority.	Exchange heuristics.	Implemented
Selim [55]	Unspecified	Assign faculty members to courses.	Assignment technique with backtracking.	Tested on real data
McClure & Wells [37]	Department of Decision Sciences, Miami University, Ohio (-,18)	Assign faculty members to courses. Respect faculty preferences.	Integer linear programming with variables representing full schedules. The number of potential schedules is curtailed.	Tested on real data
Schniederjans & Kim [54]	An unnamed department at the University of Nebraska	Assign faculty members to courses. Respect faculty preferences.	Goal programming.	Tested in real data
Dinkel, Mote & Venkataramanan [21]	Texas A&M; College of Business Administration (7,000, 300)	Assigning teachers to courses, to classrooms. Criteria: faculty preference.	Network flow (initial) Plus integer programming.	Implemented (1983-85)
Fahrion & Dollansky [25]	Department of Economics, University of Heidelberg, Germany (119,43)	Assign faculty members to courses.	Constraint logic programming.	Tested on two real data bases

Table 6. Classroom Assignment

Reference	Institution (students, courses, faculty, sections, rooms)	Problem Description	Algorithm	Results
Glassey & Mizrach [26]	University of California at Berkeley (4000,250)	Assign classes to classrooms.	6-1 linear program solved by a decomposition heuristic.	Implemented since 1985
Gosselin & Truchon [27]	Université Laval, Canada (229 to 267,57)	Assign classes to classrooms. Preferences can be expressed for some assignments.	Integer linear program solved repeatedly with different preference values. Rooms or classes can be aggregated to reduce problem size.	Tested on real data
Carter [7]	University of Waterloo (1400,120)	Assign classes to classrooms.	Integer programming with Lagrangian relaxation.	Implemented since 1986

5. Conclusion

We were somewhat surprised to discover that there are very few course timetabling papers that actually report that the methods have been implemented and used at an institution. Specifically, we found only three implementation for course timetabling, four for class teacher, three for student scheduling, three for teacher assignment and two for classroom assignment.

There were several implementations during the 1960's and 1970's using fairly simple exchange heuristics followed by a relatively quiet period in terms of research activity.

During the past ten years, there has been considerable activity and we expect to see a number of new implementations in the near future.

Acknowledgements

This work was partly supported by the Canadian Natural Sciences and Engineering Research Council under grants OGP0001359 and OGP0039782. This support is gratefully acknowledged.

References

1 Abramson, D., "Constructing School Timetables Using Simulated Annealing: Sequential and Parallel Algorithms", Man. Sci. 37, No. 1, pp. 98-113., 1991.

2 Adamidis, P. and Arapakis, P., "Weekly Lecture Timetabling with Genetic Algorithms", PATAT '97.

3 Aarts, E. and Lenstra, J.K., Local Search in Combinatorial Optimization, Wiley, Chichester, 1997.

4 Aubin, J. and Ferland, J.A., "A Large Scale Timetabling Problem", Computers & Operations Research 16, No. 1, pp. 67-77, 1989.

5 Azevedo, F. and Barahona, P., "Timetabling in Constraint Logic Programming", Proceedings of World Congress on Expert Systems'94.

6 Bloomfield, S.D. and McSharry, M.M., "Preferential Course Scheduling", Interfaces 9, No. 4, pp. 24-31, August 1979.

7 Carter, M.W., "A Lagrangian Relaxation Approach to the Classroom Assignment Problem", INFOR 27, No. 2, pp. 230-246, 1989.

8 Carter, M.W. and Tovey, C.A., "When Is the Classroom Assignment Problem Hard?", Oper. Res. 40, Supp. No. 1, pp. S28-S39, 1992.

9 Carter, M.W., Laporte, G. and Lee, S.Y., "Examination Timetabling: Algorithmic Strategies and Applications", J. of Operational Research Society 47, No. 3, 373-383, March 1996.

10 Carter, M.W. and Laporte, G., "Recent Developments in Practical Examination Timetabling", in Practice and Theory of Automated Timetabling, Springer-Verlag Lecture Notes in Computer Science 1153, Burke & Ross, eds., 1996.

11 Chahal, N. and de Werra, D., "An Interactive System for Constructing Timetables on a PC", EJOR 40, pp. 32-37, 1989.

12 Chan, H.W., Lau, C.K., and Sheung, J., "Practical School Timetabling: A Hybrid Approach Using Solution Synthesis and Iterative Repair", in Practice and Theory of Automated Timetabling, Springer-Verlag Lecture Notes in Computer Science, Burke & Carter, eds., 1998.

13 Cheng,C., Kang,L., Leung,N. and White, G.M., "Investigations of a Constraint Logic Programming Approach to University Timetabling", in Practice and Theory of Automated Timetabling, Springer-Verlag Lecture Notes in Computer Science 1153, Burke & Ross, eds., 1996.

14 Colorni A., Dorigo M. and Maniezzo V., "Genetic Algorithms - A New Approach to the Timetable Problem", Lecture Notes in Computer Science - NATO ASI Series, Vol. F 82, Combinatorial Optimization, (Akgul et al eds), Springer-Verlag, pp 235-239. 1990. (1992?)

15 Cooper, T.B. and Kingston, J.H., "The Solution of Real Instances of the Timetabling Problem", The Computer Journal, vol 36, no 7, pp 645-653, 1993.

16 Corne, D., and Ross, P., "Peckish Initialization Strategies for Evolutionary Timetabling", in Practice and Theory of Automated Timetabling, Springer-Verlag Lecture Notes in Computer Science 1153, Burke & Ross, eds., 1996.

17 Costa, D., "A Tabu Search Algorithm for Computing an Operational Timetable", European J. of Operational Res. 76, pp. 98-110, 1994.

18 Davis, L. and Ritter, L., "Schedule Optimization with Probabilistic Search", Proceedings of the 3rd IEEE Conference on Artificial Intelligence Applications, Orlando, Florida, USA, pp. 231-236, IEEE, 1987.

19 de Gans, O.B., "A Computer Timetabling System for Secondary Schools in The Netherlands", EJOR 7, 175-182, 1981.

20 de Werra, D., "Construction of School Timetables by Flow Methods", INFOR 9, No.1, 1971, 12-22.

21 Dinkel, J.J., Mote, J. and Venkataramanan, M.A., "An Efficient Decision Support System for Academic Course Scheduling", Operations Research 37, No. 6, pp. 853-864, 1989.

22 Dyer, J.S. and Mulvey, J.M., "Computerized Scheduling and Planning", New Directions for Institutional Research 13, pp. 67-86, 1977.

23 Elmohamed, S., Coddington, P., and Fox, G., "A Comparison of Annealing Techniques for Academic Course Scheduling", in Practice and Theory of Automated Timetabling, Springer-Verlag Lecture Notes in Computer Science, Burke & Carter, eds., 1998.

24 Erben,W. and Keppler, J., "A Genetic Algorithm Solving a Weekly Course Timetabling Problem", in Practice and Theory of Automated Timetabling, Springer-Verlag Lecture Notes in Computer Science 1153, Burke & Ross, eds., 1996.

25 Fahrion, R. and Dollansky, G., "Construction of University Faculty Timetables Using Logic Programming", Discrete Applied Mathematics 35, No. 3, pp. 221-236, 1992.

26 Glassey, C.R. and Mizrach, M., "A Decision Support System for Assigning Classes to Rooms", INTERFACES 16, No. 5, pp. 92-100, 1986.

27 Gosselin, K. and Truchon, M., "Allocation of Classrooms by Linear Programming", J. Operational Research Soc. 37, No. 6, 561-569, 1986.

28 Graves, R.L., Schrage, L. and Sankaran, J., "An Auction Method for Course Registration", INTERFACES 23, No. 5, pp 81-92, September-October 1993.

29 Gueret,G., Jussien,N., Boizumault,P. and Prins, C., "Building University Timetables Using Constraint Logic Programming", in Practice and Theory of Automated Timetabling, Springer-Verlag Lecture Notes in Computer Science 1153, Burke & Ross, eds., 1996.

30 Henz,M. and Wurtz,J., "Using Oz for College Timetabling", in Practice and Theory of Automated Timetabling, Springer-Verlag Lecture Notes in Computer Science 1153, Burke & Ross, eds., 1996.

31 Hertz, A., "Tabu Search for Large Scale Timetabling Problems", EJOR 54, pp. 39-47, 1991.

32 Jaffar, J. and Maher, M.J., "Constraint Logic Programming: A Survey", Journal of Logic Programming 19/20, pp. 503-581, 1994.

33 Kang, L. and White, G.M., "A Logic Approach to the Resolution of Constraints in Timetabling", European Journal of Operational Research 61, pp. 306-317, 1992.

34 Kang, L., Von Schoenberg, G.H. amd White, G.M., 1Complete University Timetabling Using Logic1, Computers and Education, 17 No. 2, pp. 145-153, 1991.

35 Lajos, G., "Complete University Modular Timetabling Using Constraint Logic Programming", in Practice and Theory of Automated Timetabling, Springer-Verlag Lecture Notes in Computer Science 1153, Burke & Ross, eds., 1996.

36 Laporte, G. and Desroches, S., "The Problem of Assigning Students to Course Sections in a Large Engineering School", Comp. & Ops. Res. 13, No. 4, pp. 387-394, 1986.

37 McClure, R.H. and Wells, C.E., "A Mathematical Programming Model for Faculty Course Assignments", Decision Sciences 15, No. 3, pp. 409-420, 1984.

38 Miyaji, I., Ohno, K. and Mine, H., 1Solution Method for Partitioning Students into Groups1, European Journal of Operational Research, 33, No. 1, pp. 82-90.616, 1981.

39 Nepal, T., Melville, S.W., and Ally, M.I., "A Brute Force and Heuristics Approach to Tertiary Timetabling", in Practice and Theory of Automated Timetabling, Springer-Verlag Lecture Notes in Computer Science, Burke & Carter, eds., 1998.

40 Ng, W.-Y., "TESS: An Interactive Support System for School Timetabling", Information Technology in Educational Management for the Schools of the Future, Fung et al (ed.), Chapman and Hall, pp. 131-137, 1997.

41 Osman, I.H. and Laporte, G., "Metaheuristics: A Bibliography. G. Laporte and I.H. Osman (eds), Metaheuristics in Combinatorial Optimization", Annals of Operations Research 63, pp. 513-623, Baltzer, Amsterdam, 1996.

42 Ostermann, R. and de Werra, D., "Some Experiments with a Timetabling System", OR Spektrum 3, pp. 199-204, 1982.

43 Paechter, B., Cumming, A. and Luchian, H., "The Use of Local Search Suggestion Lists for Improving the Solution of Timetable Problems with Evolutionary Algorithms", In Proceedings of the AISB Workshop on Evolutionary Computing, Sheffield, England, April 3-7, 1995.

44 Papoulias, D.B., "The Assignment-to-days problem in a School Time-table, a Heuristic Approach", EJOR 4, pp. 31-41, 1980.

45 Poutain, D., "Constraint Logic Programming", Byte 20, pp. 159-160, 1995.

46 Ram,V. and Scogings, C., "Automated Time Table Generation Using Multiple Context Reasoning with Truth Maintenance", in Practice and Theory of Automated Timetabling, Springer-Verlag Lecture Notes in Computer Science 1153, Burke & Ross, eds., 1996.

47 Rankin, R.C., "Automatic Timetabling in Practice", in Practice and Theory of Automated Timetabling, Springer-Verlag Lecture Notes in Computer Science 1153, Burke & Ross, eds., 1996.

48 Reeves, C.R., 1Modern Heuristic Techniques for Combinatorial Problems1, Blackwell, Oxford, 1993.

49 Rich, D.C., "A Smart Genetic Algorithm for University Timetabling", in Practice and Theory of Automated Timetabling, Springer-Verlag Lecture Notes in Computer Science 1153, Burke & Ross, eds., 1996.

50 Ross, P., Corne, D. and Fang, H.-L., "Successful Lecture Timetabling with Evolutionary Algorithms", Dept of A.I. Working Paper, University of Edinburg, 1994.

51 Sabin, G.C.W. and Winter, G.K., "The Impact of Automated Timetabling on Universities - A Case Study", J. Operational Research Soc. 37, No.7, pp. 689-693, 1986.

52 Sampson, S.E., Freeland, J.R. and Weiss, E.N., "Class Scheduling to Maximize Participant Satisfaction", Interfaces 25, No. 3, pp. 30-41, May-June 1995.

53 Schaerf, A., "Tabu Search Techniques for Large High-School Timetabling Problems", Comp. Science/Dept. of Interactive Systems, CS-R9611, Centrum voor Wiskunde en Informatica, SMC, Netherlands Organization for Scientific Researchm 1996.

54 Schniederjans, M.J. and Kim, G.C., "A Goal Programming Model to Optimize Departmental Preference in Course Assignments", Comput. Opns. Res. 14, No. 2, pp. 87-96, 1987.

55 Selim, S.M., "An Algorithm for Constructing a Unversity Faculty Timetable", Comput. Educ. 6, No. 4, pp. 323-334., 1982.

56 White, G.M., and Zhang, J., "Generating Complete University Timetables by Combining Tabu Search with Constraint Logic", in Practice and Theory of Automated Timetabling, Springer-Verlag Lecture Notes in Computer Science, Burke & Carter, eds., 1998.

57 Wright, M., "School Timetabling using Heuristic Search", Journal of the Operational Research Society, Vol. 47, pp. 347-357, 1996.

58 Yoshikawa, M., Kaneko, K., Yamanouchi, T. and Watanabe, N., "A Constraint-Based High School Scheduling System", IEEE Expert, Vol. 11, No. 1, IEEE Coputer Society, Los Alamitos, CA., pp. 63-72, February 1996.

Space Allocation: An Analysis of Higher Education Requirements

*E.K. Burke, D.B. Varley

Automated Scheduling and Planning Group
Department of Computer Science
University of Nottingham
Nottingham, UK.
ekb|dbv@cs.nott.ac.uk

Abstract. In October 1996 we sent a questionnaire on the subject of university space allocation to the estate managers of ninety six British universities. This was conducted as part of the *Automated Space Allocation* project being funded by the Joint Information Systems Committee of the UK Higher Education Funding Council. This paper describes and analyses the results obtained from these questionnaires. In particular, emphasis is placed on how large and diverse the problem is within each university, what computing tools are used if any and the constraints the universities use. We conclude by making some comments regarding about there is a requirement for an automated space allocation system and what qualities it must have.

1. Introduction

With the introduction of modularity, increasing student numbers and the continued expansion of university departments and schools, space is becoming an increasingly precious commodity. To address this, some institutions have tried to ensure efficient space utilisation by employing methods such as space charging (i.e. the charging of an amount of money, in some way or other, by the central administration for the space that each department used). However, there are a wide variety of subtle and not so subtle problems associated with space allocation and there are no rigid guidelines as to which methods should be used. We will only consider the allocation of functionality to rooms. The actual allocation of events to rooms such as examination/lecture halls has been considered separately [1]. Efficient utilisation of this type of room requires a combination of careful timetabling and space allocation. This overlap will be analysed at a later date.

The primary objective of this paper is an initial feasibility survey on the complexity and variety of the space allocation problem within UK universities. This is intended

---

* The author's names are listed in alphabetical order

to discover whether a generalised system could be created or whether the problem is too varied to make such a system feasible. The secondary objective is to obtain information regarding properties and functionality, which would be required by the universities, if the system were to be created. Thirty-eight out of ninety-six universities (40%) replied to the survey of which 14 (37%) were former polytechnics. This analysis will group all the replies together for a generalised view of this problem. However, in specific parts, such as constraint questions, comparisons will also be made between *old* and *new* UK universities to analyse the differences in approach and variety in space requirements.

There have been few papers which have addressed the problem of space allocation within academic institutions. Giannikos *et al.* [2] specifically states that academic space allocation has received relatively little attention from researchers in his paper on using goal programming for academic space allocation. Rizman *et al.* [3] also presented a goal programming model, specifically to reassign 144 offices for 289 staff members at the Ohio State University. At a lower facilities level, Benjamin *et al.* [4] used the Analytical Heirarchy Process (AHP) to determine the layout of a new computer laboratory at the University of Missouri-Rolla. The Management Information Systems Working Group of the Joint Information Systems Committee surveyed and analysed the use of computer systems within Estate and Asset Management in 1995. Unfortunately, only a brief summary of this survey is available as the statistical results are held confidentially.

2. The Space Allocation Problem

Space allocation is the process of allocating rooms or areas of space for specific functionality. It is a computationally difficult problem due to the large number of factors, which have to be simultaneously considered. Very little work has been performed in analysing the scope of this process or in developing an automated system.

In solving problems of this nature, assumptions are made about the nature of the problem, which may make finding a solution more efficient. However, when trying to create a generalised system, a major limitation is that the assumptions made must be the lowest common denominator of all possible scenarios. This survey aims to find this lowest common denominator by analysing the requirements of as many higher education institutions as possible.

2.1. Performing Space Allocation

Space allocation in a UK university is usually a large problem with many different aspects. For example, the allocation of buildings to departments is linked but separate from the process of the departments then allocating people to rooms.

According to the survey, most universities have some form of centralised office or committee controlling the global allocation of space. This office or committee allocates areas, such as buildings or floors to faculties based on requirements, current usage and utilisation figures. The faculties then allocate this space between the various departments or schools under their control. The departments then allocate specific rooms for certain functionality, such as staff offices, laboratories, etc. An exception to this rule is the centrally pooled facilities, such as lecture halls, which can be either university or externally controlled.

Due to the three-process method, a lot of data is being communicated and some replication of work is performed. However, due to the size and complexity of the problem, splitting it into these three stages is probably the most efficient way.

2.2. Different Ways of Applying Space Allocation

Space allocation is not just a single task, which is run maybe once a year. It can be applied in four different ways, some used more frequently than others:

- **Fitting all resources into a limited number of rooms**

  A department is allocated a certain number of rooms and they must allocate their resources to these rooms, fitting in all (or the majority) in the most efficient manner.

- **Minimising the number of rooms required for a set of resources**

  A department needs to be located within a building and they must calculate how many rooms they need to fully allocate all their requirements and resources.

- **Adding/removing rooms or resources from a current allocation**

  New resources or rooms must be added or removed to what is available in the best possible way, without causing too much reorganising. A balance must be made between ensuring efficient utilisation and not causing too much disruption to the resources already allocated.

- **Reorganising/optimising the current allocations**

  The department wishes to analyse the current allocation and where possible, optimise it by moving resources around within the rooms that are available.

The first two methods are used in situations when a department moves into a new building or office space. They look at the same problem from different viewpoints. The first asks how many resources can be placed within a finite number of rooms, the second asks how many rooms do we ideally need?

The third option is the most likely one to be used frequently. It allows small changes to be made to an existing allocation and tries to maintain efficient utilisation. The last option is for departments, which know they have poor space utilisation and wish to completely reorganise, regardless of the cost in moving and disruption. This option is unlikely to be used unless absolutely necessary.

2.3. Size of the Problem

The size of the space allocation problem is essentially related to the number of rooms, which need to be allocated functionality. The actual size of the universities, from small problems with 1600 rooms and 30 buildings to large problems in excess of 20000 rooms and 600 buildings, is obviously a major factor. This is often complicated by the separation of buildings over different areas or campuses. This variation in size will clearly affect the time required to perform the space allocation. It may be easy to solve a small problem optimally, whereas, it may be difficult to find a good solution for some of the larger problems. Figure 1 below shows the sizes of the universities that replied to the questionnaire in terms of rooms and buildings.

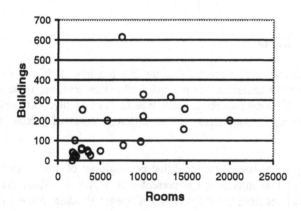

Fig. 1. Numbers of Rooms against Buildings

2.4. Reducing the Size of the Problem

As stated previously, universities are subdivided into Faculties and Departments or Schools, each usually (but not always) having control over its own space. This natural decomposition means that a large campus based problem can be divided into a series of smaller, easier problems. Figures 2 & 3 below analyses details regarding the number of rooms per subgroup. Schools are grouped together with departments for the sake of this analysis. As the graphs show, there is still a large amount of variety in the sizes of departments and faculties.

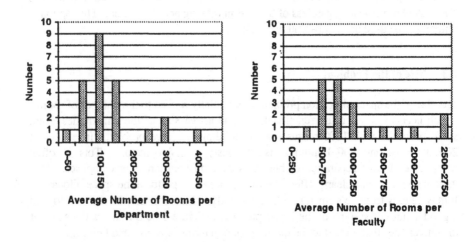

Figs. 2 & 3. Average Number of Rooms per Department & Faculty

2.5. Space Guidelines

Any process that involves allocating functionality requires guidelines to help any administrator to make decisions. In the space allocation problem, these guidelines include the sizes of rooms required for each type of functionality. Different types of staff require different room sizes, as do different kinds of lecture rooms or laboratories.

From the survey it has been discovered that no single set of guidelines is used by every university. This information is presented in Figure 4. There are different varieties of guidelines from the Polytechnics, Colleges Funding Council (PCFC) to the most common Full Time Education (FTE) 1987 guidelines. Even the FTE guideline however, is not followed rigidly with some universities modifying it based on specific requirements, historic precedence and sometimes resource limitations.

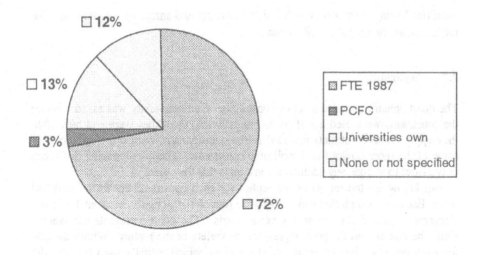

Fig. 4. Space Guidelines used by Universities in Space Allocation

2.6. Computer Usage

Few universities use computers for complete automated allocation of space. However, most institutions use computers for some parts of the process. A common situation is that universities use computers to store databases or spreadsheets which contain room/space information (e.g. room names, sizes, who is allocated to each room, what functionality each room has, etc). In some cases, Computer Aided Design (CAD) is being used to store floor plan details for each building.

Table 1 below shows how many universities are using computers in some aspect of space allocation, and for which part. These results are not mutually exclusive. Universities may use one or all of the methods.

Using computers in some aspect	97%
Computerised automation of space allocation	12%
CAD storage of building plans	62%
Database storing room information	89%
Spreadsheet storing room information	57%
Other uses	3%

Table 1. Computer Usage for Space Allocation

Various software packages are used for these methods from the most common database and spreadsheet packages on the market, to university based in-house packages. The most popular, space specific, software package used is the Archibus

Facilities Management tool, which assists in storing and analysing resource data, but has no automation for space allocation.

2.7. Constraints

The questionnaire included a set of constraints. Each university was asked whether the constraints were used and if so, to rate how important they were and how often they apply. The constraints included in the questionnaire were a few examples of some of the more obvious and applicable constraints. The questionnaire also asked universities to include any additional constraints that they used.

Shown below are the results of the replies for each constraint, shown in graphical form. Each graph is plotted with Importance (out of 10) along the horizontal axis and Occurrence (out of 100) along the vertical axis. The point represents the average while the box around the point represents the variety of the replies. Within the title for each graph is the percentage of the replies, which actually used the specific constraint within their space allocation.

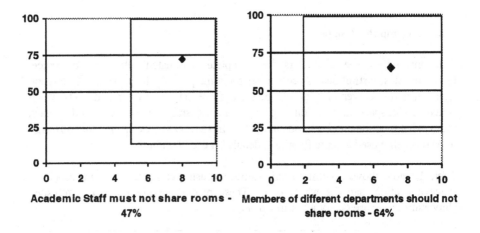

Figs. 5 & 6. Room Sharing Constraints

The above two constraints relate to which people should share rooms and when sharing should be avoided. One surprising point is that the number of universities which use the *Academic Staff Must Not Share Rooms* constraint is less than half the replies. This is further analysed in the next section where comparisons between *new* and *old* universities are discussed.

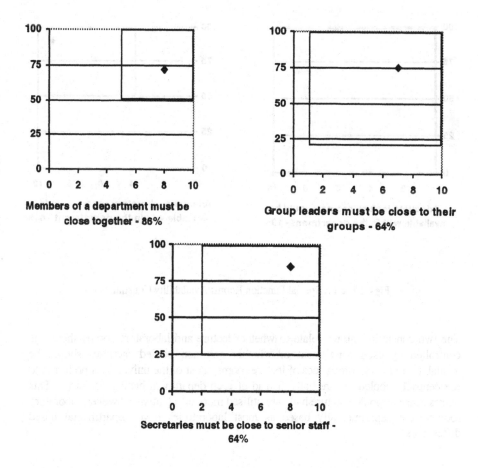

Figs. 7, 8 & 9. Locality and Positional Constraints

The three constraints shown above relate to positioning and locality of resources when allocating space. It is a common requirement to keep resources belonging to the same group together. All these constraints were used in the majority of the universities. However, only the *Keeping members of the same department together* constraint had little variation in its usage and importance.

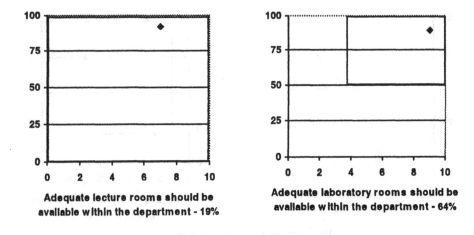

Figs. 10 & 11. Special Function Rooms Availability Constraints

The two constraints above relate to whether lecture and laboratory rooms should be controlled by each department or whether common pooled facilities should be available. In the circumstances of lecture rooms, most of the universities preferred to use centrally pooled lecture halls instead of each department having its own. This makes sense as pooled lecture halls are utilised more efficiently. However, laboratory rooms were departmentally based as most laboratories have departmental based differences.

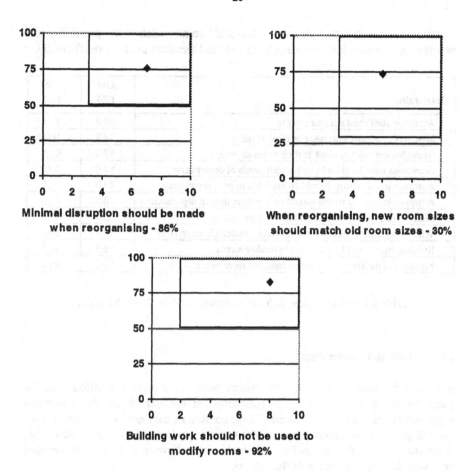

Figs. 12, 13 & 14. Reorganising Constraints

The remaining constraints relate to the reorganising and moving of resources from one allocated room to another. Most of the results for these constraints were as expected. The *Matching Room Sizes* constraint is one which has the potential for reducing the amount of work required when reorganising. This is due to the fact that recalculating the ideal amount of space required is not necessary. However, few universities used this constraint, preferring to do more work and obtain better results.

2.8. Comparison of Constraints between Old and New Universities

This section directly compares what percentage of the *old* and *new* UK universities use each constraint. The results show a slightly different approach to space allocation between the different types. The most obvious difference is in the first constraint where old universities insist on academic staff having their own rooms and many new universities seem to expect sharing. This could be attributed to the possibility that new universities have less space available and must therefore use it more efficiently.

Constraint	Old (%)	New (%)
Academic staff must not share rooms	65.2	15.4
Members of departments must be close together	78.3	100.0
Group leaders must be close to their research groups	87.0	61.5
Secretaries must be close to senior staff/heads of departments	87.0	100.0
Adequate lecture rooms should be available in each department	13.0	30.8
Adequate laboratory rooms should be available in each department	78.3	85.0
Minimal disruption should be made when reorganising	82.6	92.3
When reorganising, new room sizes should match old room sizes	30.4	30.8
Building work should not be used to modify rooms	91.3	92.3
Members of different departments must not share rooms	60.9	69.2

Table 2. Constraint Usage, Differences between Old and New Universities

2.9. Additional Constraints

A request for additional constraints, which were not explicitly mentioned in the questionnaire, was included. The following list consists of all the constraints suggested by the various institutions. These constraints may not be commonly used, but will give a good representation of the various requirements for space allocation. Some may even mutually conflict with each other, showing that some universities have very different requirements than others.

- Departments within the same faculty should be close together

- Members of staff associated with more than one department are not entitled to sole occupancy of more than one office.

- Academic staff on leave of absence for more than one term must give up their room.

- Senior academic or technical staff may have sole occupancy of a room to provide confidentiality.

- Shared buildings must have clear demarcation between sharing departments.

- Any single department, faculty or school should not be deployed over two campuses.

- Different schools and departments must be located in separate buildings.

- New space should only be allocated to departments which effectively use their current space, whose growth can be reliably predicted and whose requirements are substantiated.

- Adequate library space should be available within each department.

- Associated research groups should be located close together.

3. Conclusions

The space allocation problem within UK Higher Education institutions varies greatly. Certain constraints and requirements are common, whereas some are in direct conflict which each other. Although the process of completely allocating new resources to unused rooms will not be performed very often, space allocation is a continual process of modification as rooms and resources are gained or lost. Due to this requirement, and considering the complexity of the problem, the analysis of the survey points to there being definite scope for the construction of an automated space allocation system.

The core problem within space allocation is one of placing resources within rooms to ensure minimal space wastage. Obviously, this level of the problem is related to some standard optimisation problems such as the Knapsack problem and Bin Packing [5]. However, when we start to consider the additional constraints considered in this analysis, such as room sharing and spatial proximity groupings, the problem becomes less standard. The application and development of Meta Heuristic methods [6] are currently being investigated.

Any future generalised system must meet all the requirements specified by the universities, both those specified in this questionnaire analysis and those, which are intrinsic in any computer based system.

- **The system must have benefits for the users**

    Obviously the system must outperform (or at least match) any current methodology within an institution. It must take account of a wide variety of complex constraints and be sufficiently flexible to cope with exceptional circumstances and requirements.

- **Space allocation knowledge not computing knowledge is essential**

  A certain degree of knowledge regarding space allocation will be required for users of the system as they are required to make the final decision on whether a result is good or not. However, any system should not require the users to be computing experts. It should be intuitive and easy to use, simulating the current process as closely as possible. The users will not want to lose the overview that working through any process by hand produces. Therefore any automated system must continually provide feedback and allow changes to be made at any time. Seeing the plan progress towards completion is an integral part of space allocation from the users point of view.

- **Real test data is paramount**

  With the huge variation and scope of the space allocation problem, the only way to ensure that any future system is sufficiently generalised is to test it on as much real data as possible. Randomly created data will not provide the complexity inherent in the real world.

- **The system must cover all aspects of the problem**

  Space allocation is a complex process, involving many different aspects. From the initial allocation of areas to groups, departments and faculties, to each group allocating resources to individual rooms. Any future system must be capable of performing all these tasks as good quality space allocation relies on every stage of this process.

- **The system should not rely on data which might not be present**

  To implement the constraints in figures 5 & 6, which emphasise the requirements of resources, being located close together would require additional information regarding the spatial proximity of all the rooms available. Obviously, in large universities, if this information is unavailable then requiring the university to obtain it would be unrealistic. The system must be able to make decisions based on whatever limited information is available.

References

1. EK Burke, DG Elliman, PH Ford and RF Weare. "Exam Timetabling in British Universities – A Survey" in the *Practice and Theory of Automated Timetabling*, ed. EK Burke and P Ross, Springer-Verlag (Lecture Notes in Computer Science), 1996. Department of Computer Science, University of Nottingham, UK.

2. I Giannikos, E El-Darzi and P Lees, "An Integer Goal Programming Model to Allocate Offices to Staff in an Academic Institution" in the Journal of the Operational Research Society Vol. 46 No. 6. 713-720.

3. L Rizman, J Bradford and R Jacobs, "A Multiple Objective Approach to Space Planning for Academic Facilities" in the Journal of Management Science, Vol 25. 895-906.

4. C Benjamin, I Ehie and Y Omurtag, "Planning Facilities at the University of Missouri-Rolla". Journal of Interfaces, Vol. 22 No. 4. 95-105.

5. S Baase, "Computer Algorithms: Introduction to Design & Analysis". Second Edition, Addison-Wesley Publishing Company.

6. EK Burke, D Corne and DB Varley, "Co-Evolutionary Approaches to University Space Allocation". Proceedings of the *AISB Evolutionary Computation Conference*, Manchester, UK, 1997.

Appendix – Universities which replied

The authors are extremely grateful to all the following universities' estate managers and staff for replying to the survey :

University of Durham, Lancaster University, University of Bristol, University of Bath, University of Cambridge, University of Nottingham, Oxford University, Napier University Edinburgh, Robert Gordon Aberdeen, University of Newcastle, Royal Holloway College, University of Reading, University of Huddersfield, University of Salford, University of Wolverhampton, University of Aberdeen, University of Essex, University College London, University of Paisley Scotland, University of North London, University of Wales, College of Cardiff, University of Hull, University of Liverpool, Manchester Metropolitan University, University of Kent, University of Edinburgh, University of Birmingham, Kingston University, University of East London, Liverpool John Moores University, Keele University, University of Glamorgan, University of Stirling, Southbank University, University of Leicester, University of Bournemouth.

Tabu Search and Simulated Annealing

Off-the-Peg or Made-to-Measure?
Timetabling and Scheduling with SA and TS

Kathryn A. Dowsland

European Business Management School,
University of Wales Swansea
SA2 8PP
UK
Email: k.a. dowsland@swan.ac.uk

"A general algorithm is like a size 48 cloth. It will cover everybody but it doesn't fit any-one very well."

Abstract. Modern heuristic search techniques such as simulated annealing (SA) and tabu search (TS) are particularly suited to solving problems with a mix of hard and soft constraints or hierarchies of objectives such as those commonly encountered in real-life timetabling and scheduling problems. However, it is well-known that such methods are sensitive to the way in which the problem is modelled within a local search framework and to the generic parameters used within the algorithm. This not only raises questions concerning the robustness of a particular implementation when faced with changes in data characteristics or problem specification but also casts doubt as to the extent to which features from a solution to one family of timetabling problems may be successfully incorporated into a solution to another. This paper examines these issues from a personal point of view and uses case-studies of scheduling, timetabling and staff-rostering problems arising in the education and hospital sectors to show that it is possible to design robust solutions based on SA and TS and that lessons learned when tackling one family of problems are frequently useful for another.

1. Introduction.

Meta-heuristic search techniques such as simulated annealing and tabu search have proved to be popular and successful approaches for a variety of timetabling and scheduling problems. Initially, they were hailed as robust and flexible techniques, that were simple to use, yet capable of solving a broad range of optimisation problems. For example in their 1986 paper on simulated annealing Lundy and Mees

[1] state that "the attraction of the annealing method is that it is general yet simple to apply. Solving a problem with it requires only that one provide an adequate way of generating neighbours of solution points.". De Werra and Hertz [2] made similar claims for tabu search when they reported that "One may however wonder whether an extremely general type of search for a good solution can be developed for global optimisation problems. ...... The tabu search technique is a general method of this type." However, it is now accepted that the success of a particular implementation is sensitive to the way in which the problem is modelled within the local search framework and to the choice of parameter values within the algorithm. For many problems high quality solutions also require some sort of modification to the basic method or the incorporation of additional features, some of which may be very specific to the underlying problem structure. Thus, instead of being regarded as well-defined recipes that will work for a broad spectrum of problems, both tend to be regarded as a general framework that will, at the very least, require tuning for each new problem. These observations are particularly relevant to the area of timetabling and scheduling for a number of reasons.

First, such problems typically occur in relatively small organisations, such as schools, universities or hospitals, and although the underlying features of the problem remain constant the details will differ from institution to institution. Areas such as health and education are also high profile public services that are constantly under pressure to provide better service provision, leading to frequent changes in working practices. Thus, any implementation designed with a view to being used in more than one location, or over the medium to long term, must be sufficiently flexible to deal with these differences.

Second, apparently different problems have similar characteristics. For example, it is well known that school, course and examination scheduling problems can usually be formulated in terms of a graph-colouring model with additional objectives and constraints. At a more general level, problems across the broad spectrum of educational timetabling, staff rostering, sports scheduling etc. not only require a feasible solution subject to a set of binding constraints, but also feature a number of soft constraints or secondary objectives that should be satisfied as far as possible. In his address to the first PATAT conference Wren [3] expressed the view that, in spite of little apparent interest to date, similarities between timetabling and staff rostering may provide scope for cross-fertilisation. However, in view of the problem specific nature of successful simulated annealing and tabu search solutions, is it over optimistic to expect lessons learned in designing a solution for one problem be relevant to another?

Third, many problems have an underlying structure that can be exploited for a more efficient search. Such an approach has been adopted by Robert and Hertz [4] in their solution to constrained course scheduling problems in which they use a mixture of tabu search and specialist solution techniques for sub-problems which can be modelled as network flow and stable set problems. However it is possible that such modifications may be rendered ineffective, or even infeasible, by additional constraints or a change in objective. This view is supported by Boufflet and Nègre [5] who state that because of 'time instability', 'building a unique sophisticated method exploiting the problem characteristics is not necessarily a good idea.' This not only reinforces the notion that cross-fertilisation of ideas may not be profitable, but also raises the question of how we should deal with the trade-off between solution quality and robustness in the face of changes in problem specification.

This paper will examine these issues from the perspective of personal experience of three different timetabling and scheduling problems arising in the education and hospital sectors. The objective is not to draw any definite conclusions but to provide food for thought, and to show that at least within the limits of my own experience, the news is generally good. It *is* possible to build robust solutions based on simulated annealing and tabu search, and the knowledge gained in the process may well aid the development of solutions to new problems.

The next section contains a brief overview of simulated annealing and tabu search and their application to timetabling and manpower scheduling problems. We then go on to describe the three case studies within the following contexts:

- the robustness of method for different problem instances.
- the impact of generic parameters and problem specific information in the face of specification changes.
- the extent to which lessons learned from these or other timetabling and scheduling problems reported in the literature proved relevant in highlighting promising avenues or potential difficulties.

2. Simulated Annealing.

Simulated annealing is a variant of local search that allows some uphill moves to avoid becoming trapped early in a local optimum. Like standard local search it requires the problem to be specified in terms of a space of solutions with a neighbourhood structure defined on it, and a cost function mapping each solution onto a numeric cost or value. The process starts with any member of the solution space, usually randomly generated, and selects one of its neighbours at random. If this neighbour is better than the original it is accepted as a replacement for the current solution. If it is worse by an amount δ it is accepted with probability $\exp(-\delta/t)$, where t decreases gradually as the algorithm progresses. This process is carried out repeatedly until t is so small that no further moves are accepted, and the best solution found during the search is taken as a good approximation to the optimum. Simulated annealing was originally derived from simulations in thermodynamics and for this reason the parameter t is referred to as the temperature and the manner in which it is reduced is called the cooling schedule. An overview of the algorithm is given in Figure 1 and further details can be found in Aarts and Korst [6], Eglese [7] and Dowsland [8].

Solution quality is known to be sensitive to the way in which the problem is formulated within the local search framework - the problem specific decisions - and the cooling schedule - the generic decisions. It is generally accepted that, if sufficient time is available, long slow cooling between appropriately set limits, t_0 and t_{end} is desirable. The value of t_0 should be high enough for a large proportion of uphill moves to be accepted and t_{end} should be low enough to ensure that the system has converged. While there have been some attempts to derive a universal set of rules for defining the cooling schedule and / or to allow it to adapt automatically during the search, these have had very limited success and most implementations still use a geometric schedule i.e. reduce t according to $t \rightarrow \alpha t$ for some $\alpha < 1$ after a given number of iterations. The starting temperature may be fixed by experimentation or may be set by heating quickly until pre-determined proportion of moves are accepted. There is little formal guidance as to the problem specific decisions. However, the neighbourhood should allow any solution to be reached from any other within a finite sequence of moves and landscapes with flat plateau like areas or deep steep sided valleys should be avoided if possible.

```
Select an initial solution s0;
Select an initial temperature t0 > 0:
Select a temperature reduction function a;
Repeat
 Repeat
    Randomly select s ∈ N(s0);
    d = f(s) - f(s0);
    if d < 0 then
       s0 = s
    else
       generate random x ∈ U(0,1);
       if x < exp(-d/t) then
          s0 = s;
       endif
    endif
  until iteration_count = nrep
  t = a(t);
Until stopping condition = true.

s0 is the approximation to the optimal solution.
```

Fig. 1. The simulated annealing algorithm.

Many successful applications involve enhancements or modifications to different aspects of the search. Many of these are discussed in Dowsland [8]. Examples include non-monotonic cooling, the use of a different probability distribution, changing the neighbourhood or cost function over time and hybridisation with other methods. The success of many of these modifications depends on particular problem specific features and will not readily apply to other similar problems.

3. Tabu Search.

Like simulated annealing tabu search is a local optimisation approach that allows uphill moves. It is a more aggressive process, and in its original form it rapidly seeks out local optima by selecting the best neighbour from the neighbourhood at each iteration. As one move from the neighbourhood is always accepted tabu search will move away from any local optimum. However, without any further ingredients it would soon fall into a pattern of cycling back and fore, in and out of such an optimum. To avoid this it uses the concept of a tabu list. This defines all moves with a given attribute as tabu for a pre-specified number of iterations, known as tabu tenure. Such moves are prohibited unless the solution meets a certain aspiration criterion, usually defined as being better than the best solution so far. The attributes are chosen to prevent the search returning to a recently visited solution and should be defined by features that are easy to detect. For example in a timetabling problem in which the latest move had moved event i from timeslot j to timeslot k some possible definitions of tabu moves include;

- any move involving event i
- any move from timeslot k
- any move to timeslot j
- any move of event i to timeslot j
- any move of event i from timeslot k.

Prohibiting any one of these classes of moves will prevent the search returning to the previous solution, but they will all also eliminate different groups of other solutions, with the first three attributes being the more restrictive than the other two.

Although tabu lists are sufficient to prevent cycling they may not counter the aggression of the search sufficiently to ensure that it covers a broad enough area. Therefore a basic implementation usually includes a diversification mechanism. In its simplest form this is frequency based and penalises attributes that occur very frequently. For example, if event i has been allocated to slot j for more than p% of the last n moves then each solution with i allocated to j will be penalised to guide the search into new areas.

For a basic tabu search the problem specific decisions include those needed for simulated annealing, but also involve the tabu attributes and those to be used for frequency based diversification. The generic decisions involve the length of tabu tenure and the frequencies and penalties to be used for diversification. Many implementations also replace the search of the whole neighbourhood by searching a proportion, or even accepting the first improving move if such a move exists.

As tabu search is free of the restrictions of any physical analogy the addition of new features and exploitation of problem specific knowledge is more widespread than with simulated annealing.

Some of the best known modifications include:

- using different cost functions during the search.
- variable length tabu lists.
- candidate list strategies: these restrict the neighbourhood moves considered to those that display certain promising features.
- strategic oscillation: this forces the search to oscillate through different areas of the solution space. It is most frequently used to search the area around the feasibility boundary by repeatedly crossing in and out of the feasible region.
- ejection chains: these combine sequences of moves into chains, so that the quality of the change is measured for the chain as a whole and not for the individual components.

All of these features can dramatically improve upon a standard tabu search, but their efficient implementation frequently depends on the precise details of the problem.

4. Local Search for Timetabling and Scheduling

Local optimisation approaches are particularly attractive for timetabling and scheduling problems as most have very natural neighbourhoods. These usually involve changing the time or some other attribute of an event or swapping the attributes of two events. However, the conflict between the different objectives and constraints apparent in almost all real-life problems mean that the solution landscape is littered with local optima. Thus, although there are some algorithms based on descent strategies (e.g. Eiselt and Laporte [9]), techniques such as SA and TS that allow some uphill moves are becoming increasingly popular. Both approaches have had considerable reported success in the timetabling and scheduling arena. Examples include implementations for school timetabling problems (e.g. Abramson [10], Dige et al. [11], and Wright [12]), for examination scheduling (e.g. Thompson and Dowsland [13] and Hertz [14]), in sports scheduling (e.g. Wright [15], [16] and Costa [17]), and in manpower scheduling (e.g. Thompson [18] and Brusco and Jacobs [19]). However, the facts that almost all these examples use substantially different approaches, and that of those that have been implemented most have been at only one location, suggest that there may indeed be barriers to portability and cross-fertilisation. The following sections examine solutions to three real life scheduling problems and highlight their common features.

4.1 Computer Laboratory Scheduling.

Our first example is concerned with a solution to a laboratory scheduling problem that has been in use since 1992. At the start of each academic year in the European Business Management School at Swansea UK, several groups of between 100 and 200 students have to be allocated to computer practical sessions. For each given group of n students a laboratory of known capacity, C, is available for m sessions, where m is somewhat larger than n/C. The students concerned are studying on a variety of different degree courses, each with its own electives and there is therefore little commonality between their timetables. However, we can produce an availability matrix, A, where $a_{ij} = 1$ if student i is free to attend at time j. The problem is to produce an allocation of students to practicals so that the minimum number of timeslots i.e. $\lceil n/C \rceil$ (where $\lceil . \rceil$ denotes rounding up to the nearest integer), are utilised and each student attends at a time when he or she is available. In order to accommodate late-comers or last minute changes it is also desirable that any surplus capacity is spread out evenly over the selected sessions. After the software had been in operation for a few years an additional secondary objective was requested. This required that as far as possible a given subgroup of students be assigned together to a minimal subset of the chosen sessions, again with surplus space being spread uniformly. Full details of the solution approach can be found in Dowsland [20] and only a brief outline will be given here.

The problem was reformulated as that of finding an allocation in which each session either has no students assigned to it or is filled as closely as possible to n/C. It can then be specified within a local search framework as follows:

Feasible solutions: assignments of students to sessions when they are available.
Neighbourhood move: re-assignment of a single student.
Cost: sum of penalties over all sessions given by :

$$\sum_{j=1}^{m} g(q(j))$$

where $q(j)$ is the number of students in practical session j and

$$g(k) = \begin{cases} \sin\left(\pi \, {}^{k}\!\!\diagup\!\!{}_{\lfloor n/C \rfloor}\right) & \text{if } k \leq \lfloor n/C \rfloor \\ \sin\left(\pi \, {}^{(k-\lceil n/C \rceil)}\!\!\diagup\!\!{}_{\lfloor n/C \rfloor}\right) & \text{if } k \geq \lceil n/C \rceil \end{cases}$$

Thus each session contributes a cost function that varies with the number of students allocated as shown in Figure 2 and a zero cost solution implies all sessions are either filled to their ideal uniform occupancy or not utilised at all. This framework was used as the basis for a successful simulated annealing based decision support system that has been in use over the last five years. The details pertinent to this discussion are as follows.

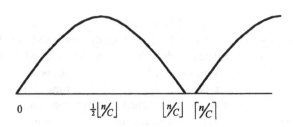

$$0 \qquad \tfrac{1}{2}\lfloor n/C \rfloor \qquad \lfloor n/C \rfloor \quad \lceil n/C \rceil$$

Fig. 2. Contribution to cost function.

A standard simulated annealing implementation consistently converged to low cost solutions but these were unacceptable in that although the majority of students were concentrated in $\lceil n/C \rceil$ sessions as required, one or two were allocated elsewhere. Closer examination revealed that the fuller sessions occupied timeslots when these 'minority' students were unavailable and that in order to escape this situation a long uphill haul would be necessary. Rather than use a frequency based diversification strategy, such as that commonly used in tabu search, to guide the search away from such unacceptable local optima in a somewhat random fashion, problem specific information was incorporated to ensure that at least one session appropriate to the 'minority' students would be fully utilised at the next local optima i.e. by ensuring that for each such student, i, the total occupancy of sessions belonging to a_i (i's row of the availability matrix) was at least $\lfloor n/C \rfloor$. This can be achieved by adding a penalty to the cost function for each unit of shortfall and then continuing the search, allowing t to increase for a while to give the search sufficient flexibility to escape to a new local optimum. If this suffered a similar problem with respect to another set of minority students the process was repeated until an acceptable, zero cost, solution was achieved.

Although the above adjustments together with an appropriate cooling schedule produced reasonable results, the algorithm was sometimes slow and occasionally failed to find a suitable solution. This was traced to biases in the sampling. Because students are available for different numbers of timeslots the most natural way of sampling - to chose a student at random and then randomly select a feasible time-slot - is biased, in that each move for a highly restricted student has a higher probability of being generated than each move for a student who can attend almost any slot. More importantly, as the probability of choosing a student from any slot is proportional to its occupancy, it also has the effect of tending to empty the fuller slots, a trend that works in the opposite direction to the objective of filling a few slots up to the level required for uniform occupancy. This problem was overcome by defining two new sampling policies designed to balance each other and using them in alternate iterations. The first randomly selects a timeslot, then a student in that slot, and then a new feasible slot for that student, while the other starts by selecting a timeslot and then selects a student free to move to that slot.

A cooling schedule that has proved robust enough to solve a number of separate problems displaying different data characteristics every year since 1992 was found without much difficulty. Reasonably fast geometric cooling using $t_0 = 0.3$ and a cooling rate of $t \rightarrow 0.8t$ is used to guide the search to good local optima. If the diversification phase is required t is allowed to rise slowly using $t \rightarrow t/(1-\beta)$ until either a new local optima is reached or $t \geq 0.9t_0$. In the latter case the schedule returns to geometric cooling, while in the former, the necessary penalties are added to the cost and slow heating is continued. In fact empirical evidence suggested that in most cases faster cooling would suffice and that for difficult problems the initial temperature and the 0.9 reheating threshold were the more critical parameters. As solutions are typically produced within a few minutes the value of 0.8 was selected to allow a margin for error.

The subgroup grouping objective was relatively easy to add to the algorithm, although it did need some additional code. This objective is simply a copy of the original applied to the specified subgroup. We therefore considered simply adding a second penalty term, in the form of a copy of the original with $q(j)$ defined over the subgroup. However, this idea was abandoned mainly due to the problems encountered in finding suitable weights. Instead the problem was tackled in phases. In our initial attempt we solved the primary problem in phase 1 and then attempted to solve the subgroup problem with a neighbourhood based on swapping a subgroup student for a non-subgroup student. This was not very successful because the subgroup problem is more difficult than the primary problem. The final solution reverses the phases, solving the subgroup problem first, and then adding the remaining students and using a neighbourhood that does not allow students from the subgroup to be reallocated.

In summary this implementation has proved sufficiently robust to provide the basis of a decision support system for allocating several different groups of practical sessions since 1992. The robustness of the cooling schedule did not present any problems but non-standard features in the form of the sampling policy, the problem specific diversification, and the use of separate phases to deal with the different objectives, all played an important part.

4.2 Examination Scheduling.

The second example concerns the problem of examination scheduling. Although our interest in the problem arose from the development of a system for University of Wales Swansea our objective was to produce a system that was flexible enough to deal with problems at other institutions. Typically this involves finding an allocation of a set of examinations to a fixed set of time slots so that no student is required to sit more than one exam at the same time (known as first order conflict), and so that the desk or room capacities are not exceeded. Most institutions have other objectives or constraints that may be binding or may simply be in the form of preferences or soft constraints. These include:

- *second order conflict* i.e. spacing out the exams for individual students by minimising the number of consecutive exams, same-day exams, or more general measures of occurrences of k exams in a given number of slots,

- *time windows* i.e. placing exams within a given time-period or in a fixed slot or placing large exams early in the exam period,
- ordering i.e. forcing pairs of exams to be held at different times, at the same time or in a given order.

Full details of our approach can be found in [13], [21] and [22]. As with the laboratory scheduling problem only those points relevant to the focus of this discussion will be outlined.

Our initial approach was to define the solution space as the set of allocations of exams to timeslots subject to the ordering constraints and to incorporate all other features into a single weighted cost function. However, we had difficulty in finding appropriate weights to ensure good, but feasible, solutions. This finding is line with our experiences with the laboratory scheduling problem and was also encountered by Dige et al. [11] in their simulated annealing approach to the school timetabling problem. In the light of its success for the laboratory scheduling problem we opted for a two phase approach with problem specific decisions as follows:

Phase 1.
Solution space: allocations of exams to time slots subject to ordering constraints
Neighbourhood move: change of time-slot for a single exam
Cost: weighted sum first-order conflict, violations of room or desk capacity and time window constraints
Starting solution: randomly generated

Phase 2.
Solution space: solutions with zero cost according to phase 1 cost function
Neighbourhood move: as phase 1 plus a neighbourhood based on the graph-theoretic concept of Kempe-chains.
Cost: weighted sum of second-order conflict and remaining objectives
Starting solution: final solution in phase 1.

Experiments with simulated annealing implementations based on their framework resulted in the following observations.

A neighbourhood based on moving a single exam produced solutions that were a great improvement on the manual approach used prior to our involvement. However, significantly better solutions were obtained using a neighbourhood based on Kempe-chains. These neighbourhoods essentially involve swapping groups of exams between two timeslots in such a way that the swap does not introduce any clashes. Experiments on data from a number of other institutions in both Europe and North America confirmed this trend for different mixes of constraints, secondary objectives, and data characteristics [22].

The natural sampling mechanisms for both neighbourhoods are non-uniform. Given the effect of biased sampling on the laboratory scheduling problem several different sampling policies were tried. Results using a variety of cost functions and data sets from different institutions confirmed that solution quality was sensitive to sampling policy, and highlighted a Kempe-chain neighbourhood sampled by selecting an exam and a new timeslot for that exam in order to fully define a single chain to be the most successful policy.

Phase 1 tended to be relatively easy to solve and a robust schedule allowing some margin for error proved fast enough for practical implementation. Phase 2 was more problematical, and on data from University of Wales Swansea used to develop and test the algorithm, a slow cooling rate starting from a relatively high starting temperature of 20 and reducing $t \rightarrow 0.99t$ every 5000 iterations, down to a stopping temperature of 0.1 was a critical factor in finding good solutions. Typical solution times for this schedule (an overnight run on a standard 486 pc) were too long to extend the range to increase robustness. We therefore conducted further experiments with the aim of finding a suitably robust schedule to allow for portability. These were based on 4 different objective functions derived from real requirements, data from 4 different institutions displaying different characteristics in terms of size, density of constraints etc., and 5 different cooling schedules, including adaptive and non-monotonic schedules from the literature. The results showed that:

1. if enough time was available then the above cooling rate, using an initial temperature at which the probability of accepting an uphill move of average size is 0.75 gave the best results. The size of an average uphill move was estimated from a small sample before commencing the search.
2. the so-called adaptive schedules tend to require problem specific parameters to perform well.
3. the problem specific decisions were robust across the spectrum of problems examined.

Thus in this case, although solution quality is sensitive to cooling, it was still possible to find a schedule that was robust enough to work well on a range of problems and in spite of differences in objectives and constraints the problem specific decisions and the two phase strategy proved robust. It is also interesting to note that the problems of biased sampling identified in the laboratory scheduling problem were also present here. Our findings also tie in with those of others working on similar problems. In particular our problem with weights was also encountered by Dige et al. [11] and Wright [12] [15] also comments on the problem of rigidly evaluating solutions as a weighted sum of penalties in his work on sports scheduling and school timetabling. He also comments that the use of Kempe-chains has enhanced his neighbourhood structure for the school timetabling problem.

4.3 Nurse Rostering.

Our interest in the third example - the nurse rostering problem - was motivated by our success in using two-phased approaches in the solution of the two previous problems. This problem, which we have been tackling for a local hospital, can be defined as follows. Each ward is manned 24 hours a day using a 3 shift system based on two day shifts and a longer night shift. For this reason a nurse on nights will tend to work less shifts in a week than if she is on days. The nurses are partitioned into three grade bands and for each shift on each day there is a minimum requirement of nurses from each band. This is expressed as a cumulative requirement i.e. at least R_1 nurses of grade 1, at least R_2 nurses of grade 2 or above and at least R_3 nurses in total. Only about half the nurses work full time, with the others on a variety of part time contracts ranging from 20% to 80% of full time hours. The rosters are worked out a week at a time, and in any one week each nurse works either days only or nights only. Thus for each nurse it is possible to define a set of feasible shift patterns, given by all combinations of the right number of days or nights. For each individual nurse in a given week some shift patterns will be preferable to others, depending on the overall quality of the pattern, whether or not it meets her off-duty requests, and her recent working history (e.g. how recently did she last work nights, or have a weekend off). All these factors can be combined into a penalty cost for each nurse shift pattern pair and the problem of finding a feasible schedule can be expressed as that of finding a solution that meets the covering requirements at all grades, while minimising the penalty costs.

Our original intention had been to tackle this problem using a two-phase approach based on simulated annealing. The first phase would find a feasible cover with enough nurses of appropriate grades on duty at all times and the second would explore the feasible region for a solution of low penalty cost. However, after an initial examination this approach was abandoned. Several sets of problem specific decisions were considered, but none of them lent themselves to a phase 2 neighbourhood that would be fast and simple to sample and yet sufficiently comprehensive to ensure an adequately connected solution space that was not divided by regions of high cost. As far as seeking out a feasible solution in phase 1 the most promising problem specific decisions appeared to be:

Solution space: allocation of each nurse to a feasible shift pattern

Neighbourhood move: change of pattern for a single nurse

Cost: shortfall in covering summed over all shifts, days and grades.

We considered including the penalty costs as a weighted term in the cost function, but initial experiments confirmed our view that this would be prone to the usual problems of balancing pressure for feasibility with the search for low penalty cost. We therefore decided to use an approach based on strategic oscillation organised in 3 cyclic phases. Phase 1 looks for a feasible solution, phase 2 explores that part of the feasible region that is reachable from this solution in the search for a solution of low penalty cost, and phase 3 re-crosses the feasibility boundary reducing penalty cost further at the expense of feasibility. The process then returns to phase 1 using a tabu list to ensure that re-entry to the feasible region is at a different point.

As a first phase in algorithm development we implemented an aggressive descent method for phase 1. Not surprisingly, with only downhill moves allowed, it usually failed to find a feasible solution. Inspired by the success of problem specific diversification in the laboratory scheduling problem we examined the features of these local optima. They fell into two classes. In the one group the reason for failure was obvious - the balance of nurses on days and nights was wrong. This can obviously be remedied by restricting allowable moves at this point to those that move a nurse in the direction required to correct the balance. The problem with the second group of optima was less obvious, but closer examination revealed that a standard tabu search might have difficulty in escaping them as they frequently formed flat-bottomed valleys with a few small humps, whereas escape would require a slightly bigger uphill step. Simply comparing the costs of neighbourhood moves would not be sufficient to guide the search in the right direction. In overcoming this problem, we incorporated problem specific knowledge that was dependent on the precise nature of the problem and developed chains of moves designed to have the overall effect of improving cover while not increasing penalty cost. The first type consists of changing the patterns worked by a chain of nurses by a single shift so that the first nurse improves the cover on a single shift and subsequent nurses in the chain cover the shift evacuated by their predecessors. In the second type each nurse works the pattern previously allocated to her predecessor with the first and last nurses chosen so that the overall effect is an improvement in the cover. Both types of chain can be generated by considering shortest paths in small graphs. They also form the basis of the neighbourhoods in phase 2.

Having made the decision to incorporate this level of problem specific information it made sense to exploit any remaining information as far as possible. The result is an effective but complex mix of candidate lists, neighbourhoods and evaluation functions, supported by two different tabu lists and a frequency based diversification strategy. The balance between aggression and diversification means that the search will rapidly seek out a whole series of good feasible local optima, and has the added advantage that different solutions of equal quality are usually produced. Full details can be found in [23].

The extent to which the final algorithm was based on the precise problem details raised the question as to what extent we had sacrificed flexibility for algorithm speed and solution quality. During the months that the algorithm was undergoing live trials we were faced with 4 changes in problem specification or working practices that go some way towards answering this question.

The first modification was the requirement that any surplus cover was to be spread uniformly throughout the weekdays. Although this was not in the original specification we had considered this as a potential requirement and it was easily dealt with by increasing the minimum cover on weekdays and introducing a dummy nurse to make up the difference between this and the surplus. The second change involved some nurses being contracted to work a mixture of days and nights in the same week. Once again this was easily accommodated, in this case by adding the additional shift patterns to the program

The third change involved the two most experienced nurses on some wards and stipulated that at most one of them should be on duty at any time on a Saturday or a Sunday. This proved more of a problem, due to the way the chain moves are generated, as care would be needed in order to make sure that chains that violated this constraint were not accepted. However, the overall effect of a chain is not known until all its moves have been considered. This problem was overcome by being overcautious and excluding a chain as soon as a problem was detected, although the problem might have been corrected later in the chain. As the restrictions apply to a subset of the potential moves for at most two nurses this did not have any noticeable effect on overall performance of the algorithm.

The last change was a result of a variation in working practice and appeared to be a significant problem. It involved a practice known as team working, in which the nurses are partitioned into two teams and at least one nurse from each team must be on duty at all times. This was considered an important objective that should override preferences and we therefore felt that it should be included as a constraint to be addressed in phase 1. However, the generation of chain moves, important to getting good solutions in phase 1 and vital to phase 2, would be rendered virtually impossible if the team constraints were to be maintained. We envisaged having to make major changes and therefore embarked on a series of experiments designed to find the best way of dealing with this problem. The first phase involved adding the team cost to the covering cost in the aggressive descent phase and simply observing the behaviour of this cost through the rest of the algorithm. The results were surprising in that although some good local optima failed to meet the team constraint the algorithm usually identified an equally good solution that did. Thus it would appear that given the power of the original algorithm in seeking out a variety of good solutions, a small nudge towards meeting the team objective, as given in the descent phase is sufficient to seek out good solutions satisfying all the requirements. This has been verified as we have not encountered any further problems with team working since the modification has been implemented.

This example illustrates that the two phase approach is not always appropriate. However, the use of problems specific knowledge to escape deep local optima was inspired by the laboratory scheduling problem, and the problem in setting appropriate weights was again in line with that experienced in any other implementations. Although we cannot draw any general conclusions from this one example, we have seen that in this case an aggressive problem specific search with respect to most of the objectives and constraints may be able to accommodate additional requirements without major changes.

5. Conclusions.

In spite of concerns about the robustness of parameters for local search algorithms such as simulated annealing and tabu search our experience has been that it is not too difficult to find parameters that will not only stand the test of time in a single location, but also cope across a spectrum of different objectives within a family of similar problems. It is also apparent that although algorithms designed for one class of problems will not be directly transferable to another, there is plenty of scope for cross-fertilisation of ideas. Thus when taken in the appropriate context lessons learned from one application area may be invaluable in trouble-shooting or searching for enhancement with another. Difficulties such as finding an appropriate way of dealing with weights are cited widely by those working in this field, while issues such as the importance of sampling featured in both our simulated annealing examples.

All three implementations proved robust to change. The examination scheduling algorithm was able to solve problems from a variety of different institutions world wide and the practical scheduling software was easily modified for the subgroup objective and has been utilised without change to deal with small differences in requirements over the years. More surprisingly the very problem specific features employed in the nurse scheduling algorithm have proved themselves amenable to four changes in problem specification, even though the last looked as though it may have caused major problems.

The evidence of these three case-studies suggests that we should not be put-off attempting to develop general solutions based on sequential search by the problem of parameter setting. Nor should worries about the trade-off between solution quality and robustness mean we necessarily shy away from exploiting problem specific features to get the most out of some of the newer features in the tabu search repertoire. Finally, we can and should learn from our own experiences and the problems and successes of others using similar approaches to timetabling and scheduling problems. In this way we can all contribute to and build on a coherent body of knowledge, that will in turn yield improved solutions for the future.

References.

1. Lundy, M. and Mees, A.: Convergence of an Annealing Algorithm. Mathematical Programming **34** (1986) 111-124
2. De Werra, D. and Hertz, A.: Tabu Search Techniques - a Tutorial and Application to Neural Networks. OR Spektrum **11** (1989) 131-141
3. Wren, A.: Scheduling, Timetabling and Rostering - a Special Relationship? In: Burke, E. and Ross, P. (eds.): Practice and Theory of Automated Timetabling, Lecture Notes in Computer Science, Vol 1153. Springer-Verlag (1996) 46-75
4. Robert, V., Hertz, A.: How to Decompose Constrained Course Scheduling Problems into Easier Assignment Type Sub-problems. In: Burke, E. and Ross, P. (eds.): Practice and Theory of Automated Timetabling, Lecture Notes in Computer Science, Vol 1153. Springer-Verlag (1996) 364-373.
5. Boufflet, J.P., Nègre, S.: Three methods used to solve an examination timetable problem. In: Burke, E. and Ross, P. (eds.): Practice and Theory of Automated Timetabling, Lecture Notes in Computer Science, Vol 1153. Springer-Verlag (1996) 327-344
6. Aarts, E., Korst, J.: Simulated Annealing and Boltzmann Machines. John Wiley and Sons (1989)
7. Eglese, R.W.: Simulated Annealing: a Tool for Operational Research. European Journal of Operational Research **46** (1990) 271-281
8. Dowsland, K.A. Simulated Annealing. In: Reeves, C.R. (ed.) Modern Heuristic Techniques for Combinatorial Problems. Blackwell (1993) 20-69.
9. Eiselt, H.A., Laporte, G.: Combinatorial Optimisation Problems with Soft and Hard Requirements. Journal of the Operational Research Society **38** (1987) 785-795.
10. Abramson, D.: Constructing School Timetables Using Simulated Annealing: Sequential and Parallel Algorithms. Management Science **37** (1991) 98-113
11. Dige, P., Lund, C., Ravn, H.F.: Timetabling by Simulated Annealing. In: Vidal, R.V.V. (ed.) Applied Simulated Annealing, Lecture Notes in Economics and Mathematical Systems, Vol 396. Springer (1993) 104-124.
12. Wright, M.: School Timetabling Using Heuristic Search. Journal of the Operational Research Society **47** (1996) 347-357

13. Thompson J.M., Dowsland, K.A.: Variants of Simulated Annealing for the Examination Timetabling Problem. Annals of Operations Research **63** (1996) 105-128

14. Hertz, A.: Tabu Search for Large Scale Timetabling Problems. *European Journal of Operational Research* **54** (1991) 39-47

15. Wright, M.: Timetabling County Cricket Fixtures Using a Form of Tabu Search. Journal of the Operational Research Society **45** (1994) 758-770

16. Wright, M.: Scheduling English Cricket Umpires. Journal of the Operational Research Society **42** (1991) 447-452.

17. Costa, D.: An Evolutionary Tabu Search Algorithm and the NHL Scheduling Problem INFOR **33** (1995) 161-178

18. Thompson, G.M.: A Simulated Annealing Heuristic for Shift Scheduling Using Non-continuously Available Employees', Computers and Operations Research **23** (1996) 275-288

19. Brusco, M.J., Jacobs, L.W.: A Simulated Annealing Approach to the Solution of Flexible Labour Scheduling Problems. Journal of the Operational Research Society **44** (1993) 1191-1200

20. Dowsland, K.A.: Using Simulated Annealing for Efficient Allocation of Students to Practical Classes. In: Vidal, R.V.V. (ed.) Applied Simulated Annealing, Lecture Notes in Economics and Mathematical Systems, Vol. 396. Springer (1993) 125-150

21. Dowsland, K.A.: Simulated Annealing Solutions for Multi-objective Scheduling and Timetabling Problems. In: Rayward-Smith, V.J., Osman, I.H., Reeves, C.R., Smith, G.D. (eds.). Modern Heuristic Search Methods John Wiley (1996) 155-167

22. Thompson, J.M., Dowsland, K.A.: General Cooling Schedules for a Simulated Annealing Based Timetabling System. In: Burke, E. and Ross, P. (eds.): Practice and Theory of Automated Timetabling, Lecture Notes in Computer Science, Vol 1153. Springer-Verlag (1996) 345-363.

23. Dowsland, K.A.: Nurse scheduling with tabu search and strategic oscillation. European Journal of Operational Research (1998) (to appear).

Generalized Assignment-Type Problems
A Powerful Modeling Scheme

Jacques A. Ferland*

Département d'informatique et de recherche opérationnelle
Université de Montréal, C.P. 6128
Succursale Centre-Ville, Montréal, Québec, Canada H3C 3J7

Abstract. Assignment-type problems and their generalized versions appear to be powerful modeling tools. In this paper we formulate several combinatorial problems and timetabling and scheduling applications as such problems. Neighborhood search techniques are very appropriate for dealing with these problems. Four different methods (descent method, Tabu search method, exchange procedure, and simulated annealing) and several restarting strategies are reviewed. Finally, we indicate how population-based techniques can be used to deal with these problems.

1 Introduction

The notion of the assignment-type problem and its generalized version appear to be powerful modeling tools. In particular, such models are very appropriate for formulating timetabling and scheduling problems where time periods have to be determined for activities according to several specific constraints. In this paper, we illustrate how this basic structure is embedded in some well-known combinatorial problems: the classical assignment problem, the generalized assignment problem, the timetabling problem, the graph coloring problem, and the graph partitioning problem. We also formulate several timetabling and scheduling applications as such problems: course timetabling, traineeship scheduling, preventive maintenance scheduling, sports league scheduling, and nurse scheduling. These illustrations are far from being exhaustive, and several other applications can be found in recent Ph.D. dissertations [4, 24]. Nevertheless, these examples should be sufficient to illustrate how useful these modeling tools are.

But in order for a model to be fully convenient, there must exist efficient solution procedures to deal with it. If a penalty approach is used to deal with the specific constraints of the model, then the feasible domain is specified only with the assignment constraints. Hence neighborhood search techniques are very appropriate heuristic methods for dealing with the transformed model since a neighborhood of a solution is easy to characterize. Four different methods (descent method, Tabu search method, exchange procedure, and simulated annealing)

---

* This research was supported by NSERC grant OGP 0008312, FCAR grant ER-1654, and FRSQ grant 930913

and several restarting (diversification) strategies are given in Section 5. Furthermore, since the feasible domain generated with the assignment constraints gives rise to a straightforward encoding scheme, genetic and hybrid genetic algorithms can be used to deal with these problems.

Once again, this paper is far from being an exhaustive survey of applications and solution procedures for the assignment-type problem. The objective is to illustrate how these modeling tools are useful for dealing with several applications. In Section 2, the notion of the assignment-type problem and its generalized version are reviewed. Several combinatorial problems and timetabling and scheduling applications are formulated as such problems in Sections 3 and 4, respectively. Some neighborhood search techniques and restarting strategies for generalized assignment-type problems are given in Section 5. Finally, in Section 6, we indicate how genetic and hybrid genetic algorithms can be used to deal with these problems.

2 Generalized Assignment-Type Problem

The *Assignment-Type Problem (ATP)* can be summarized as follows (see [9, 11]):

Given n items and m resources, the problem is to determine an assignment of each item to a resource in order to optimize an objective function and to satisfy K additional side constraints.

The mathematical model associated with an ATP is formulated as follows:

(ATP)   Min $F(x)$
              Subject to
$$G_k(x) \leq 0 \qquad 1 \leq k \leq K$$
$$x \in X(1)$$

where $X(1) = \{x : \sum_{j \in J_i} x_{ij} = 1, 1 \leq i \leq n; x_{ij} = 0 \text{ or } 1, 1 \leq i \leq n, j \in J_i\}$, and $J_i \subset \{1, 2, \ldots, m\}$ is the set of admissible resources for item i, $1 \leq i \leq n$. The decision variables x_{ij} are such that

$$x_{ij} = \begin{cases} 1 & \text{if item } i \text{ is assigned to resource } j \\ 0 & \text{otherwise.} \end{cases}$$

The objective function F and the side constraints G_k, $1 \leq k \leq K$, are only required to be calculable.

A *Generalized Assignment-Type Problem (GATP)* [7] is an ATP where each item i has to be assigned to a_i resources, $a_i \geq 1$. The mathematical model associated with a GATP is:

(GATP)   Min $F(x)$
              Subject to
$$G_k(x) \leq 0 \qquad 1 \leq k \leq K$$
$$x \in X(a)$$

where $X(a) = \{x : \sum_{j \in J_i} x_{ij} = a_i, 1 \leq i \leq n; \, x_{ij} = 0 \text{ or } 1, 1 \leq i \leq n, j \in J_i\}$, $a = [a_1, a_2, \ldots, a_n]^T$, and $a_i \geq 1$ and integer $1 \leq i \leq n$. It is easy to see that an ATP is a GATP where $a = [1, 1, \ldots, 1]^T$.

The constraints $x \in X(a)$ underlie several combinatorial problems and several timetabling and scheduling problems. In the next sections, we review some of these problems.

Furthermore, to take advantage of the simple structure of the constraints $x \in X(a)$, a *penalty approach* is used to deal with GATP. In this approach, let $V_k(x)$ denote the violation of constraint $G_k(x) \leq 0$; i.e.,

$$V_k(x) = \text{Max}\{0, G_k(x)\}.$$

Hence, for each $x \in X(a)$, let $V(x)$ denote the *total weighted violation*

$$V(x) = \sum_{k=1}^{K} \beta_k V_k(x)$$

where $\beta_k \geq 0$ is proportional to the relative importance of the kth constraint. Then, instead of solving (GATP), the solution techniques introduced in Sections 5 and 6 are applied to the following *penalized version* of GATP:

(PGATP) $$\underset{x \in X(a)}{\text{Min}} \, P(x) = \alpha F(x) + V(x)$$

where $\alpha \geq 0$. The values of α and β_k, $1 \leq k \leq K$, are parameters that can be adjusted according to the specific application. For instance, larger values for β_k, $1 \leq k \leq K$, allow putting more emphasis on feasibility.

3 Combinatorial GATP

Several combinatorial problems are instances of GATP. In this section, some of these problems are reviewed.

The first set includes three different problems where $J_i = \{1, 2, \ldots, m\}$, and the objective function F is linear:

$$F(x) = \sum_{i=1}^{n} \sum_{j=1}^{m} c_{ij} x_{ij},$$

c_{ij} being the cost of assigning item i to resource j. In the *classical assignment problem*, the side constraints are the resource assignment constraints:

$$\sum_{i=1}^{n} x_{ij} = 1 \qquad 1 \leq j \leq m,$$

and $m = n$. In the *generalized assignment problem* [23], the side constraints are the resource capacity constraints:

$$\sum_{i=1}^{n} r_{ij} x_{ij} \leq b_j \qquad 1 \leq j \leq m,$$

where r_{ij} is the amount of resource j used by item i and b_j is the amount of resource j available. Finally, in the *timetabling problem* [9, 11], the side contraints are also linear:

$$\sum_{i=1}^{n}\sum_{j=1}^{m} r_{ijk}x_{ij} \leq b_k \qquad 1 \leq k \leq K.$$

The well-known *graph coloring problem* [19] can also be seen as an ATP. Recall that the graph coloring problem is to identify a coloring for the vertices of a graph $G = (V, E)$ using a fixed number m of colors such that no two adjacent vertices have the same color. For this problem, the vertices are the items and the colors are the resources. The objective function $F(x) \equiv 0$ since the purpose is to exhibit a feasible coloring, and $J_i = \{1, 2, \ldots, m\}$. There is a side constraint associated with each color k, $1 \leq k \leq m$:

$$\sum_{(i,j)\in E} x_{ik}x_{jk} \leq 0.$$

Finally, consider the *graph partitioning problem* [21] where the purpose is to identify a partition V_1, V_2 of the set of vertices V of a graph $G = (V, E)$ such that $|V_1| = |V_2|$, which minimizes the sum of the cost of the edges with end-points in different sets. Hence, the vertices are the items and the subsets V_1 and V_2 are the two resources. The objective function is

$$F(x) = \sum_{(i,j)\in E} c_{ij}(x_{i1}x_{j2} + x_{i2}x_{j1})$$

where c_{ij} denotes the cost of edge $(i, j) \in E$. There is a side constraint associated with each subset V_k, $1 \leq k \leq 2$,

$$\sum_{i=1}^{n} x_{ik} - \frac{|V|}{2} \leq 0.$$

4   Timetabling and Scheduling GATP

In this section we review several applications in the field of timetabling and scheduling that can be formulated as GATP.

4.1   Course Timetabling [1]

Consider the problem of establishing the schedule of lectures accounting for individual student registrations and lecturer and classroom availabilities. Note that in our formulation we allow lectures of different lengths. Hence, to formulate this problem as a GATP, the lectures are the items and the starting times allowed are the resources; then for each lecture i, $1 \le i \le n$, and starting time $j \in J_i$ (the set of admissible starting times for lecture i)

$$x_{ij} = \begin{cases} 1 & \text{if lecture } i \text{ starts at } j \\ 0 & \text{otherwise.} \end{cases}$$

The objective function accounts for the lecturer preferences:

$$F(x) = \sum_{i=1}^{n} \sum_{j \in J_i} c_{ij} x_{ij}$$

where c_{ij} is the cost of starting lecture i at j specified in terms of lecturer preferences ($c_{ij} = 0$ if j is the most preferred starting time for lecture i).

A first set of side constraints is specified to eliminate the conflicting situations where students or lecturers are involved in lectures taking place simultaneously. Define:

$\Gamma_{ij} = \{(k, \ell)$: lectures i and k have students or lecturers in common, and they overlap in time if lecture i starts at j and lecture k at $\ell\}$.

The conflicting constraints are specified as follows:

$$x_{ij} x_{k\ell} \le 0 \qquad (k, \ell) \in \Gamma_{ij}, \ 1 \le i \le n, \ j \in J_i.$$

Other side constraints are introduced to account for classroom availability. Partition the set of classrooms available into several subsets including classrooms of the same type. Define

$B =$ number of classroom types
$U_b =$ number of classrooms of type b, $1 \le b \le B$.

For each lecture, a unique classroom type is specified. Define:

$K_{bt} = \{(i, j)$: lecture i requires a classroom of type b and it takes place during teaching hour t of the week whenever its starting time is $j\}$.

The classroom availability constraints are as follows:

$$\sum_{(i,j) \in K_{bt}} x_{ij} - U_b \le 0 \qquad 1 \le b \le B, \ 1 \le t \le T$$

where T is the total number of teaching hours of the week.

The course timetabling problem can be summarized as follows:

$$\text{Min} \sum_{i=1}^{n} \sum_{j \in J_i} c_{ij} x_{ij}.$$

Subject to

$$x_{ij} x_{k\ell} \leq 0 \qquad (k, \ell) \in \Gamma_{ij}, \quad 1 \leq i \leq n, \ j \in J_i$$

$$\sum_{(i,j) \in K_{bt}} x_{ij} - U_b \leq 0 \qquad 1 \leq b \leq B, \ 1 \leq t \leq T$$

$$x \in X(1).$$

The weights used to formulate the associated (PGATP) are specified to indicate the relative importance of violating the corresponding constraints. For the conflicting constraints, the weights account for the length of the conflict and the number of individuals involved. Define

$$\ell_{ijk\ell} = \text{length (in number of hours) of the conflict } (k, \ell) \in \Gamma_{ij}, \ 1 \leq i \leq n, \ j \in J_i$$

$$\delta_{ik} = \text{number of students taking both lectures } i \text{ and } k, \ 1 \leq i \leq n, \ 1 \leq k \leq n$$

$$\gamma_{ik} = \begin{cases} M + \delta_{ik} & \text{if lectures } i \text{ and } k \text{ have lecturers in common} \\ \delta_{ik} & \text{otherwise} \end{cases}$$

$$1 \leq i \leq n, 1 \leq k \leq n$$

where M is a parameter to indicate the relative importance of conflicts due to lecturers.

Let $V_{k\ell}^{ij}(x)$ denote the violation of the conflicting constraints:

$$V_{k\ell}^{ij}(x) = \text{Max}\{0, x_{ij} x_{k\ell}\} = x_{ij} x_{k\ell}.$$

Let $V_{bt}(x)$ denote the violation of the classroom availability constraints:

$$V_{bt}(x) = \text{Max}\{0, \sum_{(i,j) \in K_{bt}} x_{ij} - U_b\}.$$

Furthermore, let the weight $\beta_{k\ell}^{ij}$ associated with $V_{k\ell}^{ij}(x)$ be specified as follows:

$$\beta_{k\ell}^{ij} = \gamma_{ik} \ell_{ijk\ell}.$$

Also, let the weight β_{bt} associated with $V_{bt}(x)$ be equal to a scalar ρ for all $1 \leq b \leq B, 1 \leq t \leq T$. Then, the penalized version (PGATP) for this problem is as follows:

$$\underset{x \in X(1)}{\text{Min}} \sum_{i=1}^{n} \sum_{j \in J_i} \left[\alpha c_{ij} x_{ij} + \sum_{(k,\ell) \in \Gamma_{ij}} \gamma_{ik} \ell_{ijk\ell} x_{ij} x_{k\ell} \right]$$

$$+ \rho \sum_{b=1}^{B} \sum_{t=1}^{T} \text{Max}\left\{0, \sum_{(i,j) \in K_{bt}} x_{ij} - U_b\right\}.$$

The weight parameters α, ρ, and M are selected according to the specific context. For instance, if the priority is to reduce the conflicts, then α and ρ take smaller values.

4.2 Internship Scheduling [10]

Consider the problem of establishing the yearly schedule of students' internships. Each student has several internships (that may be of different lengths measured in terms of number of months) to complete over the year. These internships are carried out in different places. Define:

$S =$ the set of students;
$I_s =$ the set of internships of student s, $s \in S$;
$I = \bigcup_{s \in S} I_s$, the set of all internships to schedule;
$L =$ the set of places where the internships can be carried out;
$P =$ the set of one-month periods (i.e. $|P| = 12$);
$J =$ the set of place-period pairs ℓ-p, $\ell \in L$, $p \in P$;
$J_i \subset J$, the set of place-period pairs where and when internship i can start;
$T =$ the set of internship types.

To formulate this problem as a GATP, the internships are the items and the place-period pairs are the resources; then for each internship $i \in I$, and place-period pair $j \in J_i$,

$$x_{ij} = \begin{cases} 1 & \text{if internship } i \text{ starts in place-period } j \\ 0 & \text{otherwise.} \end{cases}$$

The objective function accounts for student preferences:

$$F(x) = \sum_{i \in I} \sum_{j \in J_i} c_{ij} x_{ij}$$

where c_{ij} is the cost of scheduling internship i to start in place-period j specified in terms of the preferences of the student carrying out internship i for both the place and the starting period.

A first set of side constraints is specified to eliminate the conflicting situations where a student would be carrying out more than one internship simultaneously. Define:

$\Gamma_{ij} = \{(k, \ell) : i, k \in I_s \text{ for some } s \in S, \text{ and they overlap in time if they start in place-period } j \text{ and } \ell, \text{ respectively}\}$.

The conflicting constraints:

$$x_{ij} x_{k\ell} \leq 0 \qquad (k, \ell) \in \Gamma_{ij}, \; i \in I, \; j \in J_i.$$

The places where the internships are carried out must have the supervising personnel and facilities required. Furthermore, the places may require having a minimal number of internships assigned before making the supervising personnel and facilities available. Hence define:

$\alpha_{\ell tp}$ and $\beta_{\ell tp}$ the lower and upper bounds, respectively, on the number of internships of type t that place ℓ is ready to supervise during period p, $\ell \in L$, $t \in T$, $p \in P$.

$K_{\ell tp} = \{(i,j) : i \in I, j \in J_i,$ such that i is of type t, ℓ is the location of j, and i takes place during period p if it starts in place-period $j\}$.

Now, since the minimum number ($\sum_{\ell \in L} \alpha_{\ell tp}$) of internships of type t required for all places at period p may exceed the number of such traineeships to be carried out by the set of all students, then using the bounds constraints

$$\alpha_{\ell tp} \leq \sum_{(i,j) \in K_{\ell tp}} x_{ij} \leq \beta_{\ell tp} \qquad \ell \in L, t \notin T, p \in P$$

may induce an empty feasible domain. Hence, at least for some places, we must modify the constraints to ensure that their lower bound is satisfied or that no internship is assigned there (either-or constraints). Thus the bounds constraints are replaced by

$$\alpha_{\ell tp} \leq \sum_{(i,j) \in K_{\ell tp}} x_{ij} \leq \beta_{\ell tp} \qquad (\ell, t, p) \in \bar{R}$$

$$\alpha_{\ell tp} \leq \sum_{(i,j) \in K_{\ell tp}} x_{ij} \leq \beta_{\ell tp} \text{ or } \sum_{(i,j) \in K_{\ell tp}} x_{ij} = 0 \qquad (\ell, t, p) \in R - \bar{R}$$

where $R = \{(\ell, t, p) : \ell \in L, t \in T, p \in P\}$, and $\bar{R} \subset R$ (i.e., bounds constraints are maintained for some triplets $(\ell, t, p) \in \bar{R}$, but for the other triplets, they are replaced by either-or constraints stating that the number of internships is either in the interval $[\alpha_{\ell tp}, \beta_{\ell tp}]$ or equal to 0).

But, the either-or constraints are easily transformed using additional binary variables. Indeed, since $\alpha_{\ell tp} \geq 0$, then using additional binary variables $y_{\ell tp}$ and y_r, it is easy to verify that these constraints are equivalent to

$$\alpha_{\ell tp} \leq \sum_{(i,j) \in K_{\ell tp}} x_{ij} + \beta_{\ell tp} y_{(\ell tp)(\ell tp)} \leq \beta_{\ell tp} \quad (\ell, t, p) \in R - \bar{R}$$

$$y_{(\ell tp)(\ell tp)} + y_{(\ell tp)r} = 1 \qquad (\ell, t, p) \in R - \bar{R}$$

$$y_{(\ell tp)(\ell tp)}, y_{(\ell tp)r} = 0 \text{ or } 1 \qquad (\ell, t, p) \in R - \bar{R}.$$

Furthermore, note that the additional constraints are assignment constraints. (Note that any basic text-book on operations research like [20] includes a section indicating how to transform constraints using binary variables.)

The internship scheduling problem can be summarized thus:

$$\text{Min} \sum_{i \in I} \sum_{j \in J_i} c_{ij} x_{ij}.$$

Subject to

$$x_{ij} \cdot x_{k\ell} \leq 0 \qquad (k, \ell) \in \Gamma_{ij}, i \in I, j \in J_i$$

$$\alpha_{\ell t p} \leq \sum_{(i,j) \in K_{\ell t p}} x_{ij} \leq \beta_{\ell t p} \qquad (\ell, t, p) \in \bar{R}$$

$$\alpha_{\ell t p} \leq \sum_{(i,j) \in K_{\ell t p}} x_{ij} + \beta_{\ell t p} y_{(\ell t p)(\ell t p)} \leq \beta_{\ell t p} \qquad (\ell, t, p) \in R - \bar{R}$$

$$x \in X(1), \quad y \in Y(1)$$

where

$$Y(1) = \{y : y_{(\ell t p)(\ell t p)} + y_{(\ell t p)r} = 1, \ (\ell, t, p) \in R - \bar{R}; y_{(\ell t p)(\ell t p)}, y_{(\ell t p)r} = 0 \text{ or } 1,$$
$$(\ell, t, p) \in R - \bar{R}\}.$$

In this application, we associate violation functions $V_{\ell t p}(x)$ with the bounds constraints

$$V_{\ell t p}(x) = \text{Max}\{0, \alpha_{\ell t p} - \sum_{(i,j) \in K_{\ell t p}} x_{ij}, \sum_{(i,j) \in K_{\ell t p}} x_{ij} - \beta_{\ell t p}\} \qquad (\ell, t, p) \in \bar{R},$$

rather than using two different functions associated with the lower and the upper bounds, respectively. Such a function is illustrated in Figure 4.1.

For the either-or constraints, the violation functions $V_{\ell t p}$ are

$$V_{\ell t p}(x, y) = \text{Max}\{0, \alpha_{\ell t p} - \sum_{(i,j) \in K_{\ell t p}} x_{ij} - \beta_{\ell t p} y_{(\ell t p)(\ell t p)},$$
$$\sum_{(i,j) \in K_{\ell t p}} x_{ij} + \beta_{\ell t p} y_{(\ell t p)(\ell t p)} - \beta_{\ell t p}\}(\ell, t, p) \in R - \bar{R}.$$

In Figure 4.2 we illustrate $V_{\ell t p}(x, 0)$ and $V_{\ell t p}(x, 1)$. Consider the lower envelope $V_{\ell t p}(x)$ of $V_{\ell t p}(x, 1)$ and $V_{\ell t p}(x, 0)$; i.e.

$$V_{\ell t p}(x) = \text{Min}\{V_{\ell t p}(x, 1), V_{\ell t p}(x, 0)\}.$$

With each value of $\sum_{(i,j) \in K_{\ell t p}} x_{ij}$, this function (illustrated in Figure 4.3) associates the violation of the corresponding either-or constraints when the value of $y_{\ell t p}$ is fixed in order to reduce this violation. This observation suggests the alternative of keeping the either-or constraints in the model (instead of using additional $0 - 1$ variables to replace them), and of associating with these the following violation functions:

$$V_{\ell t p}(x) = \begin{cases} \sum_{(i,j) \in K_{\ell t p}} x_{ij} & \text{if } \sum_{(i,j) \in K_{\ell t p}} x_{ij} \leq \lfloor \frac{\alpha_{\ell t p}}{2} \rfloor \\ \text{Max}\{0, \alpha_{\ell t p} - \sum_{(i,j) \in K_{\ell t p}} x_{ij}, \\ \qquad \sum_{(i,j) \in K_{\ell t p}} x_{ij} - \beta_{\ell t p}\}, & \text{otherwise} \end{cases}$$

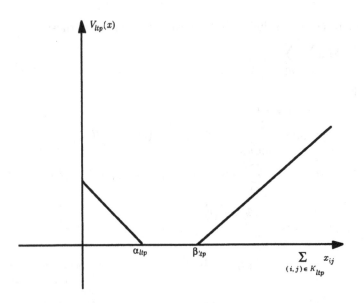

Fig. 4.1.

where $\lfloor a \rfloor$ denotes the largest integer less than or equal to a. The function illustrated in Figure 4.3 indicates that if we are closer to 0 $\left(\sum_{(i,j) \in K_{\ell t p}} x_{ij} \leq \lfloor \frac{\alpha_{\ell t p}}{2} \rfloor \right)$, then the penalty creates an advantage in moving toward 0. Otherwise, there is an advantage in moving toward $\alpha_{\ell t p}$.

4.3 Preventive Maintenance Scheduling [3]

Consider the problem of preventive maintenance scheduling for production units over a horizon of T periods of time (i.e., $1 \leq t \leq T$). Each maintenance task i ($1 \leq i \leq n$) is associated with a specific unit and lasts a specific number of periods. (Note that the same type of maintenance may be required by several units, and that several types of maintenance may be required by the same unit, but each of these instances of maintenance corresponds to a different maintenance task i.) Some of these tasks have to be repeated periodically, and the next execution must be completed not earlier and not later than a specific number of periods after the current execution. Hence several maintenance patterns j ($j \in J_i$) over the horizon are generated for each maintenance task i, and each pattern specifies the periods of the horizon during which the maintenance task i takes place; i.e., each pattern $j \in J_i$ of task i, $1 \leq i \leq n$, is characterized by a vector $\rho_{ij} = [\rho_{ij}^1, \rho_{ij}^2, \ldots, \rho_{ij}^T]$ where

$$\rho_{ij}^t = \begin{cases} 1 & \text{if task } i \text{ in pattern } j \text{ is executed during period } t \\ 0 & \text{otherwise} \end{cases}$$

$$1 \leq t \leq T.$$

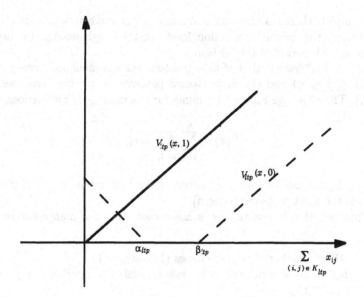

Fig. 4.2.

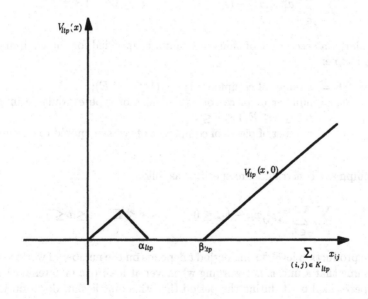

Fig. 4.3.

To complete these maintenance tasks, manpower and equipment are required. Furthermore, the overall production level must be high enough to meet the demand at each period of the horizon.

In the GATP formulation of this problem, the maintenance tasks i are the items $(1 \leq i \leq n)$ and the maintenance patterns J_i are the resources $(J = \bigcup_{i=1}^{n} J_i)$. The objective function accounts for the manager's preferences:

$$F(x) = \sum_{i=1}^{n} \sum_{j \in J_i} c_{ij} x_{ij}$$

where c_{ij} is the cost of selecting pattern j for task i (i.e., $c_{ij} = 0$ if the jth pattern is the most preferred pattern).

A first set of side constraints is associated with the manpower required. Define:

$$
\begin{aligned}
W = &\text{ number of worker classes } (1 \leq w \leq W)\\
b_{tw} = &\text{ number of workers in class } w \text{ available in period } t \ (1 \leq w \leq W,\\
&1 \leq t \leq T)\\
a_i^w = &\text{ number of workers in class } w \text{ required to execute task } i.
\end{aligned}
$$

The manpower constraints are specified as follows:

$$\sum_{i=1}^{n} \sum_{j \in J_i} a_i^w \rho_{ij}^t x_{ij} - b_{tw} \leq 0 \qquad 1 \leq w \leq W, \quad 1 \leq t \leq T.$$

Similarly, a second set of side constraints is specified for the equipment required. Define:

$$
\begin{aligned}
E = &\text{ number of equipment types } (1 \leq e \leq E);\\
\bar{b}_{te} = &\text{ number of pieces of equipement of type } e \text{ available in period}\\
&t (1 \leq e \leq E, 1 \leq t \leq T);\\
\bar{a}_i^e = &\text{ number of pieces of equipment of type } e \text{ required to execute task}\\
&i.
\end{aligned}
$$

The equipment constraints are specified as follows:

$$\sum_{i=1}^{n} \sum_{j \in J_i} \bar{a}_i^e \rho_{ij}^t x_{ij} - \bar{b}_{te} \leq 0 \qquad 1 \leq e \leq E, \quad 1 \leq t \leq T.$$

The production level during period t depends on the number of working units. We assume that a unit is not working whenever at least one maintenance task is being performed on it during the period (i.e., this unit is shut down during the period). Define:

$$
\begin{aligned}
R = &\text{ the number of units } (1 \leq r \leq R)\\
p_r = &\text{ production level of unit } r
\end{aligned}
$$

$I_r \subset \{1, 2, \ldots, n\}$ the subsets of maintenance tasks on unit r

$$(\text{i.e., } \bigcup_{r=1}^{R} I_r = \{1, 2, \ldots, n\}).$$

$U_t = $ overall production level required in period t.

The production level constraints are specified as follows:

$$\sum_{r=1}^{R} Q_r^t(x) - U_t \geq 0 \qquad 1 \leq t \leq T$$

where

$$Q_r^t(x) = \begin{cases} p_r & \text{if } \sum_{i \in I_r} \sum_{j \in J_i} \rho_{ij}^t x_{ij} = 0 \\ 0 & \text{otherwise.} \end{cases}$$

The preventive maintenance scheduling problem can be summarized thus:

$$\text{Min } \sum_{i=1}^{n} \sum_{j \in J_i} c_{ij} x_{ij}$$

Subject to

$$\sum_{i=1}^{n} \sum_{j \in J_i} a_i^w \rho_{ij}^t x_{ij} - b_{tw} \leq 0 \qquad 1 \leq w \leq W, 1 \leq t \leq T$$

$$\sum_{i=1}^{n} \sum_{j \in J_i} \bar{a}_i^e \rho_{ij}^t x_{ij} - \bar{b}_{te} \leq 0 \qquad 1 \leq e \leq E, 1 \leq t \leq T$$

$$-\sum_{r=1}^{R} Q_r^t(x) + U_t \leq 0 \qquad 1 \leq t \leq T$$

$$x \in X(1).$$

In this application, the violation functions are defined as piecewise functions to account for overtime or temporary workers, and opportunities for renting equipment. Such a function is illustrated in Figure 4.4.

4.4   Sports League Scheduling [5, 8]

This example and the nurse scheduling problem discussed in the next section are two examples of GATP where the assignment constraints $X(a)$ are such that $a \neq [1, 1, \ldots, 1]^T$.

Consider the problem of generating the schedule for a major sports league such as the National Hockey League [5, 8]. Define:

$T = $ the set of teams;

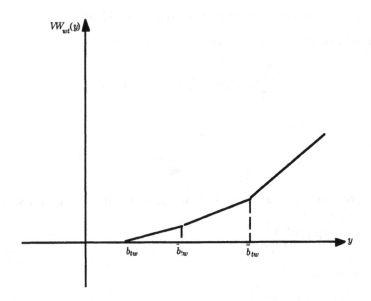

Fig. 4.4. VW_{wt} is the penalty function associated with manpower constraint w at period t, and

$$y = \sum_{i=1}^{n} \sum_{j \in J_i} a_i^w \rho_{ij}^t x_{ij}$$

$$VW_{wt}(y) = \begin{cases} 0 & \text{if } y \leq b_{tw} \\ c_1(y - b_{tw}) & \text{if } b_{tw} < y \leq \bar{b}_{tw} \\ c_1(\bar{b}_{tw} - b_{tw}) + c_2(y - \bar{b}_{tw}) & \text{if } \bar{b}_{tw} < y \leq \bar{\bar{b}}_{tw} \\ c_1(\bar{b}_{tw} - b_{tw}) + c_2(\bar{\bar{b}}_{tw} - \bar{b}_{tw}) + c_3(y - \bar{\bar{b}}_{tw}) & \text{if } y > \bar{\bar{b}}_{tw} \end{cases}$$

J = the set of dates during the season;
I = the set of game types;
n_i = the number of games of type i to be played.

A game type $i \in I$ is characterized by the visiting team t_1 and the home team t_2. Hence, the game types are the items and the dates are the resources:

$$x_{ij} = \begin{cases} 1 & \text{if a game of type } i \text{ is played on date } j \\ 0 & \text{otherwise} \end{cases}$$

for each $i \in I, j \in J_i \subset J$ where J_i is the set of dates where the arena is available for the home team of i). Then, the assignment constraints are

$$x \in X(n)$$

where $n = [n_1, n_2, \ldots, n_{|I|}]^T$.

Since the objective is to specify a feasible solution, there is no objective function, and

$$F(x) \equiv 0.$$

The number of additional side constraints increases rapidly. Furthermore, the formulation of the constraints is highly technical and requires complex notation. Hence we limit ourselves to enumerating some of the constraints in the model formulated in [8]:

- constraints forcing appropriate time delays between consecutive visits of a team to any other team;
- conflict constraints indicating that a team cannot play more than one game on a specific date;
- constraints eliminating occurrences of 2 games within 3 consecutive dates for a team;
- constraints to guarantee that each team plays at least two games each week.

4.5 Nurse Scheduling [2, 7]

Consider the problem of scheduling the working days and days off for each nurse working in a specific unit of a hospital during a specific shift over a period of time including a specified number of weeks. Define

T = the set of nurses;
J = the set of days of the period;
K = the set of weeks of the period;
J_k = the set of days of week k of the period;
n_{ik} = the number of working days of nurse $i \in I$ during week $k \in K$.

Hence, the nurses are the items and the days are the resources:

$$x_{ij} = \begin{cases} 1 & \text{if nurse } i \text{ is working during day } j \\ 0 & \text{otherwise} \end{cases}$$

for $i \in I$, $j \in J$. Then the assignment constraints are

$$x \in X(n),$$

where $n = [n_{11}, n_{12}, \ldots, n_{1|K|}, n_{21}, \ldots, n_{|I||K|}]^T$, specify that each nurse works the appropriate number of days during each week.

There is no objective function (i.e., $F(x) \equiv 0$), and as in Section 4.4, we limit ourselves to enumerating some of the additional side constraints in the model formulated in [2]:

- constraints to allow alternate weekends off for each nurse;
- constraints to limit the number of consecutive working days to an upper bound to prevent long stretches on duty;
- constraints to prevent 010 situations where a nurse has to work a single day between two days off;
- constraints to ensure the presence of the required number of supervising personnel each day;
- constraints to allow grouping accumulated holidays and extending a weekend off for a nurse;
- constraints associated with specific requests for days off during the week;
- constraints to distribute uniformly the surplus or shortage of nurses over weekdays.

In [2, 7] the authors use the multi-objective (or goal programming) approach [27] to deal with this problem where each objective corresponds to reducing the violation of a constraint. Several solution procedures are available to solve multi-objective problems, but one of them [26] is to generate a total weighted violation function $V(x)$ where the weights associated with the violation of the constraints indicate their relative priority ordering.

5 Neighborhood Search Techniques (NST)

Neighborhood search techniques [9, 23] are very appropriate for dealing with problem PGATP introduced in Section 2. NST are iterative procedures to move from a current solution $x \in X(a)$ to a neighboring solution $x' \in N(x) \subset X(a)$ until some solution $x^* \in X(a)$ is reached where x^* is acceptable according to some criterion. Now, the structure of $X(a)$ in problem PGATP allows easily characterizing the *neighborhood* $N(x)$ of any solution $x \in X(a)$. Indeed, $x' \in N(x)$ is generated from x by modifying exactly one assignment of an item to a resource; i.e., if $R(x, i)$ denotes the set of resources to which item i is assigned in solution x, $x' \in N(x)$ if there exist indices $\bar{j} \in R(x, i)$ and $j \notin R(x, i)$ such that

$$x'_{pq} = x_{pq} \qquad 1 \le p \le n \; , \; p \ne i, q \in J_p$$
$$x'_{iq} = x_{iq} \qquad q \in J_i \; , \; q \ne \bar{j}, j$$
$$x'_{i\bar{j}} = 0 \quad , \quad x'_{ij} = 1.$$

Then x' is defined as $x' = x \oplus (i, \bar{j}, j)$. It follows that $N(x) = \{z \in X(a) :$ there exists a triplet $(i, \bar{j}, j), 1 \le i \le n, \bar{j} \in R(x, i)$ and $j \notin R(x, i)$, such that $z = x \oplus (i, \bar{j}, j)\}$.

Since the size of the neighborhood $N(x)$ increases rapidly with the number of items and the number of resources, it may become prohibitive to scan the whole neighborhood to identify the best neighboring solution. Hence, in implementing NST we often restrain the search to a subset of $N(x)$ where we have a better chance of finding an improved solution. We refer to this subset as the *interesting neighborhood* (see [9]).

The different techniques are characterized by their selection process of $x' \in N(x)$ and their stopping criterion. We now summarize briefly four such techniques. The simplest one is the *descent method* which generates a local minimum. The *short-term Tabu search* allows moving away from local minima in order to search more extensively in the feasible domain. The *exchange procedure* is a descent method using a truncated enumeration tree to move away from local minima. Finally, *simulated annealing* is a probabilistic search method.

5.1 Descent Method

The descent method can be summarized as in Figure 5.1.

Initialisation:

Let $x^0 \in X(a)$ be an initial solution. $x = x^0$

General step:

Determine $x' \in N(x)$ such that $P(x') < P(x)$. If such an x' exists, then $x = x'$, and repeat the general step. If no such x' exists, then stop, and x is a local minimum.

Fig. 5.1. Descent method

5.2 Short-Term Tabu Search [15, 18]

In this technique, the solution $x' \in N(x)$ selected may cause a deterioration in the objective function (i.e., $P(x') > P(x)$) in order to be able to continue the search. Hence, a short-term Tabu list is used as a safeguard against cycling. This list TL includes the most recent modifications $(i, \bar{j}, j)$ used. Then, at each step of the method, x' is selected in $N(x) - L(x)$ where $L(x) \subset N(x)$ is the subset of neighboring solutions of x generated with the inverse of the modifications in TL. But an aspiration criterion allows selecting $x' \in L(x)$ in a case where x' improves over the best solution generated so far.

The method is summarized as follows in Figure 5.2, and the solution x' generated with the method is some kind of local minimum.

<u>Initialization:</u>

Let $x^o \in X(a)$ be an initial solution.
Let $x^* \in X(a)$ denote the best solution generated so far.
$x^* = x = x^o$.

<u>General step:</u>

Determine

$$x' \in N(x) \text{ such that } P(x') < P(x^*)$$

or

$$x' \in N(x) - L(x) \text{ such that } P(x') < P(x).$$

If no such x' exists, then determine x' such that

$$P(x') = \min_{z \in N(x) - L(x)} \{P(z)\}$$

If $P(x') < P(x^*)$, then $x^* = x'$.
$x = x'$.

Update the Tabu list TL by introducing the modification $(i, \bar{j}, j)$ used to generate x' and by eliminating the oldest modification in TL.

<u>Stopping criteria</u>

Stop either

    after a maximal number of iterations

or

    after a maximal number of successive iterations with no improvement.

Fig. 5.2. Short-Term Tabu Search

5.3 Exchange Procedure [11]

This NST can be seen as a sequence of successive completions of the descent method starting with an initial solution generated from the current local minimum. To be more specific, once a local minimum x^* is identified at the end of the descent method, a new initial solution is generated to restart the descent method by applying a truncated depth-first enumeration tree search where x^* is the root of the tree. The search is truncated to monitor the number of modifications (i.e., the depth) used and the level of deterioration allowed along the way before identifying a new solution x such that $P(x) < P(x*)$. (Note that each branch of the tree is associated with a modification. Hence, the end-nodes of a branch are neighbors of each other.) Then, the descent method is reapplied with the initial solution x.

This procedure is in the spirit of the Tabu search method since it allows a deterioration in the objective function in order to move away from a local minimum. The major difference is the monitoring of the process to ensure an improvement in the objective function before reapplying the descent method.

The stopping criteria are similar to those in the Tabu search method, but the procedure can also fail (i.e., it stops) to identify a new improving initial solution to restart the descent method. Note that in this case, an alternative is to modify the depth or the level of deterioration in order to allow the enumeration process to continue.

5.4 Simulated Annealing [21]

In the simulated annealing technique, x' is selected randomly in $N(x)$, and it becomes the current solution (i.e., $x = x'$) according to a probability decreasing with the value of $(P(x') - P(x))$ (the deterioration of the objective function) and with the time elapsed since initialization.

Initialization:

Let $x^0 \in X(a)$ be an initial solution.
Let $x^* \in X(a)$ denote the best solution generated so far.
$x^* = x = x^0$.

General step:

Determine randomly $x' \in N(x)$.
If $P(x') \leq P(x)$, then
$-\ x = x'$
$-$ if $P(x) < P(x^*)$, then $x^* = x$.

If $P(x') > P(x)$, then
$-$ select a random number $r \in [0,1]$
$-$ if $r \leq e^{-\frac{P(x')-P(x)}{TP}}$, then $x = x'$.

Stopping criteria

Stop either

   after a maximal number of iterations

or

   after a maximal number of successive values of TP with no improvement.

Fig. 5.3. Simulated Annealing

The technique can be summarized as in Figure 5.3. Note that the value of TP (called the temperature) decreases periodically during the procedure in order to reduce the tolerance for deterioration as time elapsed since initialization increases. The second stopping criterion indicates that the procedure stops whenever x^* is not improved and the proportion of time that the current solution is changed in too small for a maximal number of successive values of TP (see [9, 21]).

5.5 Diversification

In order to search more extensively the feasible domain $X(a)$ and to increase our chances of identifying a better local minimum, Glover and Laguna [16] suggest using diversification strategies to restart the NST with new initial solutions. Such strategies are introduced in [22] for the quadratic assignment problem, but they are easily adapted for GATP. They can be summarized as follows:

i) Random diversification.

A new initial solution is generated by assigning the items to the resources randomly.

ii) First order diversification.

In this strategy, we use only information related to the current local minimum x^* to generate the new initial solution. This iterative procedure starts with x^* as the initial solution, and it moves further away from x^* at each iteraction by eliminating an assignment found in x^*.

To introduce the procedure, denote $N(x, x^*) = \{z \in X(a) :$ there exists a triplet $(i, \bar{j}, j), 1 \le i \le n, \bar{j} \in R(x, i) \cap R(x^*, i)$ and $j \notin R(x, i) \cup R(x^*, i)$, such that $z = x \oplus (i, \bar{j}, j)\}$, the subset of elements in the neighborhood of x generated by eliminating an assignment found in x^*. Then the procedure is summarized as in Figure 5.4.

iii) Second order diversification.

In this strategy, we use the cumulated information related to all local minima generated so far with the NST. The new initial solution is generated by assigning items to resources to which they have been assigned least often in the local minima generated so far.

Initialisation:

Let $x = x^*$.

General step:

Determine $x' \in N(x, x^*)$ such that

$$P(x') = \text{Min}_{z \in N(x, x^*)}\{P(z)\}$$

$x = x'$.

Stopping criteria:

Stop either
after a maximal number of iterations
or
when $P(x) < P(x^*)$.

Fig. 5.4. First order diversification

5.6 Solving the Applications

In this section we indicate some references where the applications in Section 4 are solved with NST. This is summarized in Table 5.1

	Techniques		
Applications	Tabu Search	Exchange Procedure	Diversification
Course Timetabling	[4]	[1]	
Traineeship Scheduling	[10]		[10]
Maintenance Scheduling	[3]	[3]	
Sports League Scheduling	[5]	[8]	
Nurse Scheduling	[2, 7]		[7]

Table 5.1.

All the applications were solved using short-term Tabu seach techniques. The technique used in [2, 7] for the nurse scheduling problem is a variant of Tabu search adapted for goal programming where the values of the solutions are compared lexicographically (see [7] for more details).

Both Tabu search and the exchange procedure are used in [3] to solve the preventive maintenance scheduling problem for power generating units. The numerical tests on several variants of the problem indicate that the techniques are competitive.

Finally, diversification strategies were implemented in [7, 10]. The tests involve comparing a long Tabu search (with a large number of iterations and a large number of successive iterations with no improvement) with the three diversification strategies using a short Tabu search as NST. As expected, in general the quality of the solution is better when a diversification strategy is used, but the solution time also increases. Furthermore, the first order diversification strategy seems to generate results of higher quality than the others, but it also requires more time.

6 Genetic Algorithms and Hybrids

Genetic algorithms [6, 17] are population-based procedures where the population (a subset of feasible solutions $x \in X(a)$) is transformed by means inspired by evolution and natural selection. An encoding scheme is required to associate an individual of the population with each solution. For ATP, the natural encoding is to associate with each solution a vector

$$z = [z_1, z_2, \ldots, z_n]^T$$

where z_i is the index of the resource to which item i is assigned. Similarly, for GATP, each solution can be characterized by a vector

$$z = [z^{1^T}, z^{2^T}, \ldots, z^{n^T}]^T$$

where $z^i \in \mathbb{R}^{a_i}$ is a vector including the indices of the resources to which item i is assigned.

Once the encoding scheme is specified, a specific genetic algorithm is characterized by

i) a selection mechanism for individuals of the population (parents) that will be used for reproduction;
ii) a reproduction mechanism (crossover operator) to produce offspring from selected parents;
iii) a mutation mechanism (mutation operator) that can be used to modify some individuals of the population;
iv) a culling scheme to manage how to population is maintened and which individuals are removed from the population.

Hence, several algorithms can be specified according to the different mechanisms used.

Our experience in using genetic algorithms to deal with several combinatorial problems (quadratic assignment problem [14], graph coloring [12, 13], maximum clique [12], and satisfiability [12]) indicates that they are very time-consuming. Hence, genetic hybrid algorithms [12, 13, 14] have been used to accelerate these algorithms for the combinatorial problems mentioned above and to improve the quality of the solutions obtained. In a nutshell, these hybrids can be described as genetic algorithms where an NST is used as the mutation mechanism which is systematically applied to each offspring generated. These hybrids are still very time-consuming, but they have the advantage of generating solutions of very high quality for instances of these problems which are difficult to solve (see [12]).

7 Conclusion

In this paper, we formulate several combinatorial problems and several timetabling and sceduling problems in the framework of generalized assignment type problem. The purpose is to illustrate how to use the framework, rather than giving an exhaustive survey of such applications. Furthermore, four different neighborhood search techniques are briefly overviewed together with three different approaches to restart these (i.e., to introduce diversification in the search). Finally, we indicate how genetic algorithms can be used to deal with generalized assignment type problems.

References

1. Aubin, J., Ferland, J.A.: A Large Scale Timetabling Problem. Computers and Operations Research **16** (1989), 67–77.
2. Berrada, I., Ferland, J.A., Michelon, P.: A Multi-Objective Approach to Nurse Scheduling with both Hard and Soft Constraints. Socio-Economic Planning Science **30** (1996), 183–193.

3. Charest, M., Ferland, J.A.: Preventive Maintenance Scheduling of Power Generating Units. Annals of Operations Research **41** (1993), 185–206.

4. Costa, D.: Méthodes de résolution constructives, séquentielles et évolutives pour des problèmes d'affectations sous contraintes. Doctoral dissertation, Department of Mathematics, École Polytechnique Fédérale de Lausanne, Lausanne, Switzerland (1995).

5. Costa, D.: An Evolutionary Tabu Search Algorithm and the NHL Scheduling Problem. INFOR **33** (1995), 161–178.

6. Davis, L.: Handbook of Genetic Algorithms. Van Nostrand Reinhold, New York (1991).

7. Ferland, J.A., Berrada, I., Nabli, I., Ahiod, B., Michelon, P., Gascon, V.: Generalized Assignment-Type Goal Programming Problem and Application to Nurse Scheduling. Publication #1112, département d'informatique et de recherche opérationnelle, Université de Montréal, Montréal, Canada (1998).

8. Ferland, J.A., Fleurent, C.: Computer Aided Scheduling for a Sport League. INFOR **29** (1991), 14–25.

9. Ferland, J.A., Hertz, A., Lavoie, A.: An Object-Oriented Methodology for Solving Assignment-Type Problems with Neighborhood Search Techniques. Operations Research **44** (1996), 347–359.

10. Ferland, J.A., Ichoua, S., Lavoie, A., Gagné, E.: Scheduling Medical School Students' Internships Using Tabu Search Methods with Intensification and Diversification. Publication #1068, département d'informatique et de recherche opérationnelle, Université de Montréal, Montréal, Canada (1998).

11. Ferland, J.A., Lavoie, A.: Exchange Procedures for Timetabling Problems. Discrete Applied Mathematics **35** (1992), 237–253.

12. Fleurent, C., Ferland, J.A.: Object-Oriented Implementation of Heuristic Search Methods for Graph Coloring, Maximum Clique, and Satisfiability. DIMACS Series in Discrete Mathematics and Theoretical Computer Science **26** (1996), 619–652.

13. Fleurent, C., Ferland, J.A.: Genetic and Hybrid Algorithms for Graph Coloring. Annals of Operations Research **63** (1996), 437–461.

14. Fleurent, C., Ferland, J.A.: Genetic Hybrids for the Quadratic Assignment Problem. DIMACS Series in Discrete Mathematics and Theoretical Computer Science **16** (1994), 173–187.

15. Glover, F.: Future Paths for Integer Programming and Links to Artificial Intelligence. Computer and Operations Research **13** (1986), 533–549.

16. Glover, F. and Laguna, M.: Tabu Search. In: Reeves, C. (ed): Combinatorial Problems. Blackwell Scientific Publishing, New-York (1993), 70–150.

17. Goldberg, D.E.: Genetic Algorithms in Search, Optimization, and Machine Learning. Addison-Wesley, Reading Massachusetts (1989).

18. Hansen, P.: The Steepest Ascent, Mildest Descent Heuristic for Combinatorial Programming. Congress on Numerical Methods in Combinatorial Optimization, Capri, Italy (1986).

19. Hertz, A., DeWerra, D.: Using Tabu Search Techniques for Graph Coloring. Computing **39** (1987), 345–351.

20. Hillier, F.S., Lieberman, G.J.: Introduction to Operations Resarch. 5th edn. McGraw-Hill, New York (1990).

21. Johnson, D.S., Aragon, C.R., McGeoch, L.A., Shevon, C.: Optimisation by Simulated Annealing: An Experimental Evaluation; Part I, Graph Partitioning. Operations Research **37** (1989), 865–892.

22. Kelly, J.P., Lagana, M., Glover, F.: A Study of Diversification Strategies for the Quadratic Assignment Problem. Computers and Operations Research **21** (1994), 885–893.

23. Reeves, C. (ed.): Modern Heuristic Techniques for Combinatorial Problems. Blackwell Scientific Publishing, New York (1993).

24. Robert, V.: La confection d'horaires par décomposition en sous-problèmes d'affectation. Doctoral dissertation, Department of Mathematics, École Polytechnique Fédérale de Lausanne, Lausanne, Switzerland (1996).

25. Ross, C.T., Soland, R.M.: A Branch and Bound Algorithm for the Generalized Assignment Problem. Mathematical Programming **3** (1975), 91–103.

26. Sherali, H.D.: Equivalent Weights for Lexicographic Multi-Objective Programs: Characterizations and Computations. European Journal of Operational Research **11** (1981), 367–379.

27. Steuer, R.E.: Multiple Criteria Optimization: Theory, Computation, and Application. Krieger, Malabar Florida (1989).

An Examination Scheduling Model to Maximize Students' Study Time*

Bernd Bullnheimer

Department of Management Science, University of Vienna
Bruenner Str. 72, A - 1210 Vienna, Austria
bernd.bullnheimer@univie.ac.at

Abstract. In this paper we develop a model for small scale examination scheduling. We formulate a quadratic assignment problem and then transform it into a quadratic semi assignment problem. The objective of our model is to maximize student's study time as opposed to minimizing some cost function as suggested in other QAP approaches. We use simulated annealing to demonstrate the model's ability to generate schedules that satisfy student as well as university expectations. Furthermore the application of the model to a real world situation is presented.

1 Introduction

The examination scheduling problem is to assign a number of exams to a number of potential time periods or slots within the examination period taking into account that no student can take two or more exams at the same time, as well as several other constraints (cf. [7], p.4).

The most common way of modelling the basic examination scheduling problem is as a graph coloring problem (cf. e.g. [23]). In such an approach the nodes of the graph represent the exams. Two "exam nodes" of the graph are connected by an edge if there is at least one student taking both exams. The chromatic number of the corresponding graph gives the minimum number of colors it takes to color the vertices of the graph. It is also the minimum number of time periods needed to generate a clash free exam schedule for exams with the same "color" can be scheduled at the same time without causing a clash[1]. The graph coloring problem is NP-complete (cf. [16]).

Besides the first-order conflicts there are so-called back-to-back or second-order conflicts (cf. [3]). This terms a situation where a student has to take two consecutive exams. Such situations put extra strain on students and should therefore be avoided. If weekends and other breaks of the examination period are not considered, this minimization problem is reduced to the well-known traveling salesman problem which is also NP-complete (cf. [13]).

---

* We would like to thank Richard Hartl, Asoka Wöhrmann and three anonymous referees for their helpful comments.
[1] A situation, where a student would have to take two or more exams at a time is called a clash, a first-order conflict or a direct conflict.

Finally, there may be further constraints dealing with room capacities, pre-scheduled exams, exams that have to be scheduled in a predefined order and so-called higher-order conflicts (cf. [8]). That is, students having something like two exams within 24 hours or three exams in two days. To deal with these higher-order conflicts several authors (cf. e.g. [1], [2], [12], [20], [24]) have used a quadratic assignment problem (QAP) formulation, which again is NP-complete.

The examination scheduling problem is a highly complex combinatorial problem consisting of NP-complete subproblems. Usually acceptable solutions to this problem are conflict free regarding first-order conflicts (and other constraints that are considered hard constraints by the institutional authorities) and are "not too bad" regarding second- and higher-order conflicts (soft constraints)[2]. On the other hand there are also approaches minimizing the number of first-order conflicts[3]. Problems considered in literature vary in size depending on the degree of centralization between university wide (e.g. [4], [5]) and nation wide (cf. [14]) coordination and are tackled by using all kind of different approaches[4].

In this paper we present a small scale model for a faculty wide examination timetabling problem which does not only avoid clashes but also maximizes students' study time. In addition to that it gives the university authorities the possibility to allocate their resources to other tasks by reducing the number of required time periods. The proposed model has (ex-post) been applied to a real world examination scheduling problem. Solutions generated by a simulated annealing algorithm are presented and discussed. The remainder of the paper is organized as follows: in Sect. 2 the model is developed. Sect. 3 shows its application and the resulting schedules. Finally, we conclude and give an outlook to future research in Sect. 4.

2 Model

2.1 Basic Formulation – QAP

Considering only small scale examination scheduling problems, we assume there are T slots or time periods and also T exams to be scheduled[5]. Therefore the examination scheduling problem can be stated as a quadratic assignment problem as follows:

$$\sum_{s=1}^{T}\sum_{t=1}^{T}\sum_{p=1}^{T-1}\sum_{q=p+1}^{T} n_{pq}d_{st}y_{ps}y_{qt} \quad \rightarrow \quad \max \quad (1)$$

s.t.

$$\sum_{t=1}^{T} y_{pt} = 1 \quad \forall p \in \{1, ..., T\} \quad (2)$$

---

[2] Cf. e.g. [8], [9], [24].

[3] Cf. e.g. [15], [21].

[4] Overviews are given in [6] and [7].

[5] This assumption might not be valid for larger universities. If there are more exams than slots, the exams have to be grouped first as proposed in e.g. [20], [24].

$$\sum_{p=1}^{T} y_{pt} = 1 \qquad \forall t \in \{1, ..., T\} \tag{3}$$

$$y_{pt} \in \{0, 1\} \qquad \forall p, t \in \{1, ..., T\}$$

where

$$y_{pt} = \begin{cases} 1 & \text{if exam } p \text{ is scheduled at time period } t \\ 0 & \text{otherwise} \end{cases}$$

n_{pq} number of students to take exams p and q

d_{st} "distance" between time period s and time period t

The main difference between our approach and other QAP approaches is the following: instead of introducing cost values for assigning exams p and q to slots s and t, and minimizing the objective function (as done in e.g. [12], [19] and [20]) we define the time between two slots as "distance". Under the assumption that students can use this time to study, the overall study time for all students is then maximized by (1). Constraints (2) and (3) ensure that each exam is assigned to exactly one time slot and vice versa. In this context, the time between the beginning of the examination period and the first exam a student takes is irrelevant and can therefore be omitted[6]. At this point it should be stated that most ways of ensuring fairness in this context are questionable. All possible cost values represent the price for extra strain done to a student and the overall *"welfare"* is maximized, what might yield hard consequences for some students. The idea of minimizing the maximum strain put on any student is possible but not common. This approach would only lead to a rearrangement of the exams that could be done in a more straightforward way by using a traveling salesman formulation. The advantage of our quadratic model will be shown in the next section where we transform it into a quadratic semi assignment problem (QSAP) by relaxation of some constraints.

2.2 Advanced Formulation – QSAP

In the basic model clashes cannot occur because each exam is assigned to another time period. In many situations this approach may not be practicable in terms of university resource allocation as more capacity (room and personnel) than necessary is allocated. For that reason there should be the possibility to assign more than one exam to a given slot. Therefore we relax the corresponding constraint (3) and formulate a QSAP. To avoid first-order conflicts, we introduce

---

[6] To be exact, the sum of the time periods between any two exams a student has to take, aggregated over all students is maximized. Problems arising from possible double countings caused by students taking four or more exams (which also occur in cost minimizing approaches) are neglected.

a penalty in the objective function by setting $d_{ss} = -M$ for all time periods with M being a very large number. At the same time we add a fixed cost term for every time period actually used for at least one exam and state the problem as follows:

$$\sum_{s=1}^{T} \sum_{t=1}^{T} \sum_{p=1}^{T-1} \sum_{q=p+1}^{T} n_{pq} d_{st} y_{ps} y_{qt} \quad - \sum_{t=1}^{T} f_t u_t \quad \rightarrow \quad \max \quad (4)$$

s.t.

$$\sum_{t=1}^{T} y_{pt} = 1 \qquad \forall p \in \{1, ..., T\} \qquad (5)$$

$$\sum_{p=1}^{T} y_{pt} \le T u_t \qquad \forall t \in \{1, ..., T\} \qquad (6)$$

$$u_t, y_{pt} \in \{0, 1\} \qquad \forall p, t \in \{1, ..., T\}$$

where

$$u_t = \begin{cases} 1 & \text{if at least one exam is scheduled at time period } t \\ 0 & \text{otherwise} \end{cases}$$

f_t \quad fixed cost if at least one exam is scheduled at time period t

Now (6) ensures that the maximum number of exams scheduled at any time period is T, if the fixed cost for that time period is "paid", and 0 otherwise. This formulation (4)–(6) is rather flexible and adjustable to the specific situation of a certain university[7]. The cost value f_t can be set equal to a constant F if the university authorities are indifferent about what slots remain "free". If the examination period should be short, increasing the cost term for later slots leads to solutions with only the early slots used. One way of doing this is to penalize only the last slot actually used, denoted by T_{max}. Multiplied by a constant cost term F this results in an increasing penalty FT_{max}.

At this point it is clear, that these increasing penalties transform the problem into a multiobjective problem. The students' objective is to have the exams spread as widely as possible while university authorities prefer schedules with

---

[7] Capacity constraints regarding the maximium number of students that can take an exam at a given time period can be stated by replacing (6) with:

$$\sum_{p=1}^{T} n_p y_{pt} \le c_t u_t \qquad \forall t \in \{1, ..., T\}$$

n_p = number of students to take exam p

c_t = maximium number of students that can take an exam at time period t

fewer time periods required. This criterion conflicts with the students' objective (cf. [7], p.5). When constant cost terms are used the problem still has two objectives, but they are not necessarily conflicting.

As mentioned earlier, the problems discussed in literature differ when it comes to second- and higher-order conflicts. In most cases however, avoiding back-to-back conflicts has priority. Maximizing students' study time might not sufficiently spread the exams to avoid higher-order conflicts, as one additional day of study time for five students adds as much to the objective function as five additional days for one student. For those reasons we set the distance for all slots that are back-to-back equal to zero and furthermore introduce a parameter $\alpha \in [0, 1]$ to modify the distance matrix as follows:

$$\tilde{d}_{st} = \begin{cases} -M & \text{if } s = t \\ 0 & \text{if } s \text{ and } t \text{ are "back-to-back"} \\ (d_{st})^\alpha & \text{otherwise} \end{cases}$$

The parameter α again enables the institution to give more or less emphasis on the spread of the exams. Setting $\alpha = 0$ means that only back-to-back conflicts are minimized and no further attention is given to the amount of time a student can study between two exams as long as they are not back-to-back. On the other hand, $0 < \alpha < 1$ implies that the "marginal product of study time" is positive but diminishing and finally, $\alpha = 1$ is the other extreme and implies that the "marginal product of study time" is constant. Thus, choosing an appropriate value for the parameter α on which university authorities as well as students can agree, replaces the task of finding penalty values for the conflicts *clash, consecutive, sameday, back-to-back across night*, and *within 24 hours* that are used in eg. [12].

Before presenting the application to a real world examination timetabling problem, the two formulations of the model actually used are summarized. In both cases the objective is to maximize students' study time. At the same time, in (7)-(9), which we call the model with constant cost, the number of slots used is minimized:

$$\sum_{s=1}^{T} \sum_{t=1}^{T} \sum_{p=1}^{T-1} \sum_{q=p+1}^{T} n_{pq} \tilde{d}_{st} y_{ps} y_{qt} - \sum_{t=1}^{T} F u_t \to \max \tag{7}$$

s.t.

$$\sum_{t=1}^{T} y_{pt} = 1 \qquad \forall p \in \{1, ..., T\} \tag{8}$$

$$\sum_{p=1}^{T} y_{pt} \leq T u_t \qquad \forall t \in \{1, ..., T\} \tag{9}$$

$$y_{pt}, u_t \in \{0, 1\} \qquad \forall p, t \in \{1, ..., T\}$$

On the other hand, the duration of the exam period is additionally minimized in (10)-(13), which we call the model with increasing cost. There (13) ensures that T_{max} can not be less than t, if at least one exam is assigned to that time period t:

$$\sum_{s=1}^{T}\sum_{t=1}^{T}\sum_{p=1}^{T-1}\sum_{q=p+1}^{T} n_{pq}\tilde{d}_{st}y_{ps}y_{qt} - FT_{max} \to \max \qquad (10)$$

s.t.

$$\sum_{t=1}^{T} y_{pt} = 1 \qquad \forall p \in \{1, ..., T\} \qquad (11)$$

$$\sum_{p=1}^{T} y_{pt} \le Tu_t \qquad \forall t \in \{1, ..., T\} \qquad (12)$$

$$tu_t \le T_{max} \qquad \forall t \in \{1, ..., T\} \qquad (13)$$

$$y_{pt}, u_t \in \{0, 1\} \qquad \forall p, t \in \{1, ..., T\}$$

In the next chapter the presented versions of the model are applied to a real world examination timetabling problem and the results obtained are discussed.

3 Application and Results

3.1 Problem Description

During winterterm 1994/1995 the newly founded Faculty of Economics and Management at the Otto-von-Guericke-University Magdeburg offered 15 "Grundstudium" courses[8] and 27 "Hauptstudium" courses[9]. The corresponding exams took place in a three-week period (15 working days) at the end of the term. As students can only take either Grund- or Hauptstudium courses, the problem described can be divided into two independent problems: one problem with 391 students taking 15 exams (1737 sittings), and another problem with 419 students taking 27 exams (1682 sittings). Room capacities were sufficiently large so they were not restrictive and could be neglected. As the authorities in charge for the exam schedule did not know about examination timetabling, a quite basic schedule that consisted of 3 slots per day (at 8am, 10am and 3pm) was used. To rule out clashes only one exam was assigned to each slot. The 15 Grundstudium exams were scheduled on a daily basis (using the 10am slots) and the 27 Hauptstudium exams were scheduled 2 per day (using the 8am and 3pm slots) with 3 days having only one exam in the afternoon.

The work presented in this paper is an ex-post optimization of the described problem. We show to what extent both parties involved, students and university

---

[8] Comparable to undergraduate courses.
[9] Comparable to graduate courses.

authorities could benefit by allowing more than one exam to be assigned to any slot. Better schedules could be generated without changing the given time structure of the schedule, i.e. the slots[10].

3.2 Local Search Approach

The aim of this study was neither to find the problem's optimal solution nor the best possible algorithm to solve it, but to show that the proposed model combined with a local search algorithm is capable to produce good solutions. For reasons of complexity it was obvious that only a heuristic approach would be appropriate, even though the problem under consideration was relatively small. With simulated annealing (cf. [17]) a powerful but easy to implement local search procedure, that has been successfully used for other timetabling problems (cf. eg. [22]), was chosen. The idea of simulated annealing is to repeatedly choose a solution S' within the neighbourhood of the current solution S and compare the quality of S and S'. Higher quality solutions are always accepted whereas lower quality solutions are accepted with the following probability:

$$P_{Accept(S')} = exp(\frac{Quality(S) - Quality(S')}{T})$$

where the *temperature* T controls how likely it is to accept a lower quality solution[11].

Our neighbourhood structure was defined by two classes of movements: *exam moves* and *slot moves*. Exam moves alter the current solution as follows: one exam is randomly picked and scheduled to another randomly picked slot, no matter whether this slot is used or not. An example for an exam move is shown in Fig. 1, where (randomly picked) exam F, previously assigned to slot 7 in solution S, is assigned to (randomly picked) slot 6 in solution S', together with exam C which was already scheduled for that slot.

S	A	E, H			G	B	C	F	D
S'	A	E, H			G	B	C, F		D

Fig. 1. Example for an exam move

Slot moves on the other hand, do not work on single exams but rather on complete slot assignments (all exams assigned to a slot), changing the order of these previously made slot assignments. Thereby the possibilities are a) two exchange, b) shift of a sequence, c) inversion of a sequence, and d) random

---

[10] That explains why we could assume that the number of slots is equal to the number of exams (cf. Footnote 5).

[11] For more details on simulated annealing the reader is referred to e.g. [18].

rearrangement of a sequence. Examples for these moves are illustrated by Fig. 2. The slot move "two exchange" in Fig. 2a) works as follows: Two slots (here slots 2 and 5) are randomly picked and all exams scheduled for the first one (exams E and H assigned to slot 2 in S) are assigned to the second one (slot 5 in S') and vice versa.

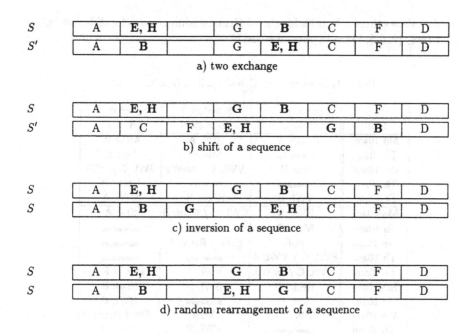

a) two exchange

b) shift of a sequence

c) inversion of a sequence

d) random rearrangement of a sequence

Fig. 2. Examples for a slot move

For the slot moves on basis of a sequence, shown in Fig. 2 b) – d), two random numbers are needed to define the sequence: the first slot of the sequence (here slot 2) and the length of the sequence (here 4 slots). All slot assignments of such a sequence are then randomly shifted (here by 2 slots), inverted or randomly rearranged (cf. Fig. 2).

All results presented in this study are the best solutions found after 5000 iterations of the simulated annealing algorithm. The initial temperature was set to $T_0 = 10$ and was reduced by a cooling factor $\delta = 0.975$ every time a move was accepted[12]. In each iteration, the current solution was altered by applying one of the four different slot moves described above, where each had a 25% chance to be selected. After that, an exam move was applied with a certain probability, that was initially set to 25% and then gradually decreased to 2% during the

---

[12] Due to the stated aim of the study not much emphasis was put on the tuning (cooling schedule, number of iterations etc.) of the simulated annealing algorithm. Instead, a parameter setting that has been found to work well was used.

search. The quality of the resulting neighbourhood solution was then evaluated by using (7) and (10) respectively and accepted according to the acceptance criterion mentioned above.

3.3 Application of the Model with Constant Cost

Table 1 shows three schedules for the 15 Grundstudium exams for different values of α ($\alpha \in \{0, 0.5, 1\}$) with constant cost[13].

Table 1. Schedules for Grundstudium (constant cost)

Slot	Schedule 1 $\alpha = 0$	Schedule 2 $\alpha = 0.5$	Schedule 3 $\alpha = 1$
Mo 10am	Math A	BWL A	Math A
Tu 10am	▬▬▬	Stat B	Stat B
We 10am	Stat B	VWL C / ReWe	BWL C / VGR
Th 10am	▬▬▬	Stat A	Recht B
Fr 10am	VWL B	Recht B	VWL A
Mo 10am	BWL A	BWL C / Math B	Stat A
Tu 10am	Math B	▬▬▬	▬▬▬
We 10am	VGR	EDV / Recht A	▬▬▬
Th 10am	BWL C / VWL A	▬▬▬	▬▬▬
Fr 10am	VWL C / ReWe	VWL A	VWL C / ReWe
Mo 10am	Recht B	BWL B	Math B
Tu 10am	Stat A	▬▬▬	BWL B
We 10am	BWL B	VGR	EDV / Recht A
Th 10am	▬▬▬	VWL B	VWL B
Fr 10am	EDV / Recht A	Math A	BWL A
back-to-back conflicts	23	166	260

In all cases three slots remain unused, i.e. a fifth of the total examination period could have been saved. For $\alpha = 0$ a solution with only 23 back-to-back conflicts was found. Increasing the value of α tends to shift the free slots towards the center of the examination period. The black bars indicating unused slots, are scattered over the examination period when $\alpha = 0$ and centered when $\alpha = 1$ (cf. Table 1). At the same time increasing α causes the largest exams Math A and BWL A (which also have the highest number of students taking both exams, namely 189) to move to the first and last slot respectively. Simultaneously the relative importance of back-to-back conflicts diminishes and their number grows to 166 ($\alpha = 0.5$) and finally to 260 ($\alpha = 1$).

---

[13] Objective function values are not reported as they are distorted by different values of α.

For the Hauptstudium the potential savings are even larger. Schedules using only 10 out of 27 time periods can be generated. Again, for $\alpha = 1$ the free slots are concentrated in the middle of the exam period and all second week slots could have been saved for the "price" of 468 back-to-back conflicts. With $\alpha = 0$ they could be brought down to at least 61[14]. Both schedules are presented in Table 2.

Table 2. Schedules for Hauptstudium (constant cost)

Slot	Schedule 1 $\alpha = 0$	Schedule 2 $\alpha = 1$
Mo 8am	Industrieök.	Wachstumsth.
Mo 3pm	Theorie Rent.	Marketing / Geldth.
Tu 3pm	▬▬▬	Personalm.
We 8am	Steuern IV / Dyn.Opt.	Fin.I / Prod.Wirt.I / Psy.Gdl.
We 3pm	Umweltökon.	Planungsmod.
Th 3pm	▬▬▬	Umweltökon.
Fr 8am	Kostenrech. / Ber.Lern.	Steuern IV
Fr 3pm	Plan.mod. / Geldth. / Berufserz.	Konz. / Dyn.Opt. / Geldpol.
Mo 8am	Berufspäd. / Arbeitsr.	▬▬▬
Mo 3pm	▬▬▬	▬▬▬
Tu 8am	▬▬▬	▬▬▬
Tu 3pm	Steuern I / C & P	▬▬▬
We 8am	▬▬▬	▬▬▬
We 3pm	▬▬▬	▬▬▬
Th 8am	Marktforschg.	▬▬▬
Th 3pm	Konz.	▬▬▬
Fr 8am	▬▬▬	▬▬▬
Fr 3pm	Fin.I / Prod.Wirt.I / Psy.Gdl.	▬▬▬
Mo 8am	Marketing / Geldpolitik	Lagerh. / NLP / Ber. Lern.
Mo 3pm	▬▬▬	Industrieök.
Tu 8am	▬▬▬	Kostenrech.
Tu 3pm	NLP / Wachstumsth.	Termingesch.
We 8am	Personalm.	Theorie Rent. / Berufserz.
We 3pm	▬▬▬	Organisat.
Th 3pm	Lagerh.	Marktforschg.
Fr 8am	Termingesch.	C & P / Berufspäd. / Arbeitsr.
Fr 3pm	Organisat.	Steuern I
back-to-back conflicts	61	468

---

[14] This solution is definitely not optimal as the two Thursday exams in week two (Marktforschg. and Konz.) have seven students in common and therefore at least seven back-to-back conflicts could be avoided by shifting Marktforschg. to the Wednesday afternoon slot.

3.4 Application of the Model with Increasing Cost

The application of the model with increasing cost had similar effects: a low α leads to schedules with few back-to-back conflicts and, on the other hand, a high α again shifts the largest exams to the first and last slot. The difference now is that no matter what value of α was used the unused slots were actually the last slots of the exam period[15]. Table 3 shows two solutions for the Grundstudium problem generated with α set equal to zero and equal to one.

Table 3. Schedules for Grundstudium (increasing cost)

Slot	Schedule 1 $\alpha = 0$	Schedule 2 $\alpha = 1$
Mo 10am	Stat B	BWL A
Tu 10am	Stat A	Stat B
We 10am	BWL A	VWL C / VGR
Th 10am	Math B	BWL C / Recht B
Fr 10am	EDV / Recht A	Math B
Mo 10am	VWL B	Stat A
Tu 10am	ReWe	VWL A
We 10am	BWL C / Recht B	ReWe
Th 10am	VWL A	BWL B
Fr 10am	Math A	EDV / Recht A
Mo 10am	BWL B	VWL B
Tu 10am	VWL C / VGR	Math A
We 10am	▬▬	▬▬
Th 10am	▬▬	▬▬
Fr 10am	▬▬	▬▬
back-to-back conflicts	79	234

3.5 Showing Cost Independence

The results show that allowing more than one exam per time period is beneficial because in the best solutions found, actually more than one exam is assigned to some slots. There is however an obvious trade-off between student and university interests in the increasing cost model: shortening the examination period always cuts the potential study time for some students. The exam situation at the Otto-von-Guericke University Magdeburg was not that tight and therefore university authorities had no reason to promote a model that students would probably not agree on. But even for the constant cost model students argued that they would have to "pay the price" for schedule improvement caused by unused time periods and extremely high values for F.

---

[15] The value of F was chosen relatively high compared to the number of students.

To show that the results do not depend on the fixed cost and still hold, even if there is no reward for an unused slot, F was set equal to zero in (7). As a benchmark, the simulated annealing algorithm was run for four different α–values ($\alpha \in \{0, 0.5, 0.75, 1\}$) with a random one-exam-per-slot assignment as initial solution. To ensure that every schedule had exactly one exam assigned to every time period, only slot moves were performed. The resulting objective function values of the best solutions found after 5000 iterations are shown in the rows "exams per slot = 1" in Table 4 for the Grundstudium- and the Hauptstudium problem respectively[16].

Table 4. Objective function values for $F = 0$

Grundstudium				
exams per slot	$\alpha = 0$	$\alpha = 0.5$	$\alpha = 0.75$	$\alpha = 1$
= 1	3344	9003	15928	29142
≥ 1	3354	9005	15959	29267

Hauptstudium				
exams per slot	$\alpha = 0$	$\alpha = 0.5$	$\alpha = 0.75$	$\alpha = 1$
= 1	2862	7575	13444	24754
≥ 1	2866	7655	13575	25352

Then simulated annealing was run again, but this time slot as well as exam moves were performed. The corresponding objective function values (shown in the rows "exams per slot ≥ 1") are in all cases higher, i.e. these schedules give students more time to study. In addition to that, in all schedules some slots had parallel exams and consequently some slots remained unused. They only differed with respect to which time periods had exams assigned to them: for low α-values the free slots were scattered over the exam period whereas a high α had a tendency to center the free slots. Therefore the option to assign more than one exam to a time period is not only better from the students' point of view but also preferable in terms of university resource allocation[17].

4 Conclusion

In this paper we present a model suitable for small scale examination timetabling problems, that takes student as well as university matters into consideration. At

---

[16] Remember that comparing schedules for different values of α on the basis of objective function values is misleading as they depend highly on α.

[17] The "more than one exam per slot allowed" problem is a relaxation of the "only one exam per slot" problem and therefore no worse solutions could be expected. But it was essential to show students that the extra flexibility in the problem was advantageous for them and that the two different objectives were not conflicting.

the same time it gives much flexibility to university authorities, and enables them to reduce the number of time periods required and to shorten the exam period respectively. The appropriate use of the parameters α and F in either the constant or the increasing cost model will control the form of the schedules generated. Using slightly different values for α, a set of schedules can easily be generated and university authorities should have the expert knowledge to choose the most satisfactory schedule from that set. In this way the decision about the actual examination timetable is supported.

The model is also applicable to larger problems if some kind of grouping (using e.g. graph colouring) of the exams is performed first. Then it is advantageous, that the proposed algorithm can still alter the grouping (what might lead to better schedules) and therefore not too much emphasis has to be put on the grouping algorithm. The approach using a quadratic formulation together with the control parameter α proved to be promising as the difficult task of finding adequate penalties for all kinds of conflicts is unnecessary. Furthermore the maximization of the study time seems to lead to more fairness. Suppose university authorities set F to a reasonable value the presented model supports them in responsibly allocating scarce resources.

Subject of future research should be the inclusion of further constraints, the tuning of the simulated annealing algorithm as well as a study on the use of other metaheuristics (e.g. ant colony optimization [10]) and the quality of their results. Furthermore an application to seminar timetabling [11] is planned.

References

1. **Arani, T., M. Karwan and V. Lofti (1988):** A Langrangian relaxation approach to solve the second phase of the exam scheduling problem, *European Journal of Operational Research* 34, 372-383.
2. **Arani, T. and V. Lofti (1989):** A Three Phased Approach to Final Exam Scheduling, *IIE Transactions* 21, 86-96.
3. **Balakrishnan, N., A. Lucena and R.T. Wong (1992):** Scheduling examinations to reduce second-order conflicts, *Computers and Operations Research* 19, 353-361.
4. **Boufflet, J.P. and S. Negre (1996):** Three Methods Used to Solve an Examination Timetable Problem, in: Burke, E. and P. Ross (eds.): *Practice and Theory of Automated Timetabling*, Springer, Berlin, 327-344.
5. **Burke, E.K., J.P. Newall and R.F. Weare (1996):** A Memetic Algorithm for University Exam Timetabling, in: Burke, E. and P. Ross (eds.): *Practice and Theory of Automated Timetabling*, Springer, Berlin, 241-250.
6. **Carter, M.W. (1986):** A survey of practical applications of examination timetabling algorithms, *Operations Research* 34, 193-202.
7. **Carter, M.W. and G. Laporte (1996):** Recent Developments in Practical Examination Timetabling, in: Burke, E. and P. Ross (eds.): *Practice and Theory of Automated Timetabling*, Springer, Berlin, 3-21.
8. **Carter, M.W., G. Laporte and J.W. Chinneck (1994):** A General Examination Scheduling System, *Interfaces* 24, 109-120.
9. **Desroches, S., G. Laporte and J.-M. Rousseau (1978):** HOREX: a computer program for the construction of examination schedules, *INFOR* 16, 294-298.

10. **Dorigo, M., Maniezzo, V. and A. Colorni (1996):** Ant System: Optimization by a Colony of Cooperating Agents. *IEEE Transactions on Systems, Man, and Cybernetics* 26 (1), 29-41.

11. **Eglese, R.W. and G.K. Rand (1987):** Conference Seminar Timetabling, *Journal of the Operational Research Society* 38 (7), 591-598.

12. **Ergül, A. (1995):** GA-Based Examination Scheduling Experience at Middle East Technical University, in: Burke, E. and P. Ross (eds.): *Practice and Theory of Automated Timetabling*, Springer, Berlin, 212-226.

13. **Garey, M.R. and D.S. Johnson (1979):** *Computers and Intractability: A Guide to the Theory of NP-Completeness*, W.H.Freeman, New York.

14. **Hansen, M.P., V. Lauersen and R.V.V. Vidal (1995):** Nationwide Scheduling of Examinations: Lessons from Experience, paper presented at: *First International Conference on the Practice and Theory of Automated Timetabling*, Edinburgh, 30th Aug. - 1st Sept. 1995.

15. **Johnson, D. (1990):** Timetabling university examinations, *Journal of the Operational Research Society* 41, 39-47.

16. **Karp, R.M. (1972):** Reducibility among Combinatorial Problems, in: Miller, R.E. and J.W. Thatcher (eds.): *Complexity of Computer Computations*, Plenum Press, New York, 85-103.

17. **Kirkpatrick, S., C.D. Gelatt and M.P. Vecchi (1983):** Optimization by Simulated Annealing, *Science* 220, 671-680.

18. **Laarhoven, P.J.M. van and E.H.L. Aarts (1988):** *Simulated Annealing: Theory and Applications*, Kluwer, Dordrecht.

19. **Laporte, G. and S. Desroches (1984):** Examination Timetabling by Computer, *Computers and Operations Research* 11 (4), 351-360.

20. **Leong, T.Y. and W.Y. Yeong (1987):** Examination Scheduling: A Quadratic Assignment Perspective, *Proc. Int. Conf. on Optimization: Techniques and Applications*, Singapore, April 1987, 550-558.

21. **Mehta, N.K. (1981):** The application of a graph coloring method to an examination scheduling problem, *Interfaces* 11, 57-64.

22. **Thompson, J. and K.A. Dowsland (1996):** General Cooling Schedules for a Simulated Annealing Based Timetabling System, in: Burke, E. and P. Ross (eds.): *Practice and Theory of Automated Timetabling*, Springer, Berlin, 345-363.

23. **Welsh, D.J.A. and M.B. Powell (1967):** An upper bound for the chromatic number of a graph and its application to timetabling problems, *Computer Journal* 10, 85-86.

24. **White, G.M. and P.W. Chan (1979):** Towards the construction of optimal examination schedules, *INFOR* 17, 219-229.

A Comparison of Annealing Techniques for Academic Course Scheduling

M. A. Saleh Elmohamed[1], Paul Coddington[2], and Geoffrey Fox[1]

[1] Northeast Parallel Architectures Center
Syracuse University, Syracuse, NY 13244, USA
{saleh, gcf}@npac.syr.edu
[2] Department of Computer Science
University of Adelaide, S.A. 5005, Australia
paulc@cs.adelaide.edu.au

Abstract. In this study we have tackled the NP-hard problem of academic class scheduling (or timetabling) at the university level. We have investigated a variety of approaches based on simulated annealing, including mean-field annealing, simulated annealing with three different cooling schedules, and the use of a rule-based preprocessor to provide a good initial solution for annealing. The best results were obtained using simulated annealing with adaptive cooling and reheating as a function of cost, and a rule-based preprocessor. This approach enabled us to obtain valid schedules for the timetabling problem for a large university, using a complex cost function that includes student preferences. None of the other methods were able to provide a complete valid schedule.

1 Introduction

The primary objective of this study is to derive an approximate solution to the problem of university class scheduling, or timetabling, which can be summarized as follows: given data sets of classes and their days, enrollments, and instructors; rooms and their capacities, types, and locations; distances between buildings; priorities of each building for different departments; and students and their class preferences; the problem is to construct a feasible class schedule satisfying all the hard constraints and minimizing the medium and soft constraints. Hard constraints are space and time constraints that must be satisfied, such as scheduling only one class at a time for any teacher, student, or classroom. Medium and soft constraints are student and teacher preferences that should be satisfied if possible.

The timetabling problem (TTP) is a high-dimensional, non-Euclidean, multi-constraint combinatorial optimization problem, and is consequently very difficult to solve. This problem has been tackled by many researchers, mostly in the field of operations research. A number of different heuristics have been tried on different instances of the problem, from high school to university course scheduling (see the reviews by de Werra [5] and Shaerf [30] and the papers collected in Ref. [4]). For small to medium size problems, such as exam scheduling, high school scheduling, or course scheduling for a university department, many of these methods work well. However no particular method has yet been shown to produce good results for real-world problems on a much larger scale, such as scheduling all courses for a large university, which we address in this paper. Also, we are not aware of any large scale study that takes into account constraints due to student preferences, as we have done.

We have used data for classes at Syracuse University. Currently this problem is handled by the university scheduling department in a semi-automated fashion. A scheduling program is used to find a partial solution, and substantial manual effort is required to iterate towards a final solution. Also, when scheduling a certain semester (e.g. fall 1996), a template of a previous semester (e.g. fall 1995) is used as part of the input data.

We have applied the following optimization techniques to this problem:

1. A rule-based expert system.
2. Mean-field annealing.
3. Simulated annealing with geometric cooling.
4. Simulated annealing with adaptive cooling.
5. Simulated annealing with adaptive cooling and reheating as a function of cost.
6. Simulated annealing (using each of the three different cooling schedules) with a rule-based preprocessor to provide a good initial solution.

The best results were obtained using simulated annealing with adaptive cooling and reheating as a function of cost, and with a rule-based preprocessor to provide a good initial solution. Using this method, and with careful selection of parameters and update moves, we were able to generate solutions to the class scheduling problem using real data for a large university. None of the other methods were able to provide a complete valid solution.

2 The Timetabling Problem

Timetabling is the assignment of time slots to a set of events, subject to constraints on these assignments. The NP-complete *professors and classes* timetabling problem [7, 13, 14] is a constraint satisfaction problem that can be briefly stated as follows:

For a certain school with N_p professors, N_q classes, N_x classrooms and lecture halls, and N_s students, it is required to schedule N_l professor-class pairs within a time limit of N_t time slots producing a legal schedule. A legal schedule needs to be found such that no professor, class, or student is in more than one place at a time, and no room is expected to accommodate more than one lesson at a time or more students than its capacity.

The constraints for this problem can be hard, medium or soft. The medium and soft constraints have an associated cost (or penalty), and if they are not satisfied, the goal is to minimize this cost. Soft constraints have a lower priority (and thus lower cost) than medium constraints. The hard constraints must be satisfied, so their associated cost must be reduced to zero. A feasible schedule is one that satisfies all the hard constraints.

Hard constraints are usually constraints that physically cannot be violated. This includes events that must not overlap in time, such as:

- classes taught by the same professor,
- classes held in the same room,
- a class and a recitation or a lab of the same class.

Another examples are space or room constraints:

- A class cannot be assigned to a particular room unless the capacity of the room is greater than or equal to the class enrollment.
- Some classes, such as laboratories, require a certain type of room.

Medium constraints are usually considered to be those constraints that fall into the gray area between the hard and soft constraints [9]. In our implementation, we define medium constraints to be constraints such as time and space conflicts which, like hard constraints, cannot physically be violated (for example, it is not possible for one person to be in two different classes at the same time). However we consider these constraints to be medium rather than hard if they can be avoided by making adjustments to the specification of the problem. The primary example is student preferences. We cannot expect to be able to satisfy all student class preferences, in some cases, certain students will have to adjust their preferences since certain classes will clash, or will be oversubscribed.

Medium constraints have a high penalty attached to them, although not as high as that associated with the hard constraints. In the final schedule the penalty of these constraints should be minimized and preferably reduced to zero. Some examples of medium constraints are:

- Avoid time conflicts for classes with students in common.
- Eligibility criteria for the class must be met.
- Do not enroll athletes in classes that conflict with their sport practice time (of course, depending on the sport).

Soft constraints are preferences that do not deal with time conflicts, and have a lower penalty (or cost) associated with them. We aim to minimize the cost, but do not expect to be able to reduce it to zero. Some examples are:

- For each student, balance the three-day (*Mon, Wed, Fri*) as well as the two-day (*Tue, Thu*) schedules.
- Balance or spread out the lectures over the week.
- Classes may request contiguous time slots.
- Balance enrollment in multi-section classes.
- Lunch and other break times may be specified.
- Professors may request periods in which their classes are not taught.
- Professors may have preferences for specific rooms or types of rooms.
- Minimize the distance between the room where the class is assigned and the building housing its home department.

Some soft constraints may have higher priority (and thus higher cost) than others. For example, preferences involving teachers will have higher priority than the preferences of students.

The **cost function** measures the quality of the current schedule and generally involves the weighted sum of penalties associated with different types of constraint violations. The aim of the optimization technique is to minimize the cost function.

3 Mean-Field Annealing

One of the potential drawbacks of using simulated annealing for hard optimization problems is that finding a good solution can often take an unacceptably long time. Mean-field annealing (MFA) attempts to avoid this problem by using a deterministic approximation to simulated annealing, by attempting to average over the statistics of the annealing process. The result is improved execution speed at the expense of solution quality. Although not strictly a continuous descent technique, MFA is closely related to the Hopfield neural network [15, 17].

Mean-field annealing has been successfully applied to high school class scheduling [14]. For scheduling, it is advantageous to use a Potts neural encoding to specify discrete neural variables (or neurons) for the problem. This is defined in its simplest form as a mapping of events onto space-time slots, for example an event i, in this case a professor-class pair (p, q), is mapped onto a space-time slot a, in this case a classroom-timeslot pair (x, t). Now, the Potts neurons S_{ia} are defined to be 1 if event i takes place in space-time slot a, and 0 otherwise. In this way, the constraints involved can be embedded in the neural net in terms of the weights $w_{i,j}$ of the neural network, which encode a Potts normalization condition such as $\sum_a S_{ia} = 1$.

For a full derivation of the mean-field annealing algorithm from its roots in statistical physics, see Hertz *et al.* [15] or Peterson *et al.* [29]. Here we will just give a brief overview of the method. The basic idea is that it is possible to approximate the actual cost or energy function E, which is a function of discrete neural variables S_{ia}, by an effective energy function E' that can be represented in terms of continuous variables U_{ia} and V_{ia}. These are known as mean field variables, since V_{ia} is an approximation to the average value of S_{ia} at a given temperature T.

This approach effectively smooths out the energy function and makes it easier to find the minimum value, which is obtained by solving the saddle point equations $\frac{\partial E'}{\partial V_{ia}} = 0$ and $\frac{\partial E'}{\partial U_{ia}} = 0$, which generate a set of self-consistent mean field theory (MFT) equations in terms of the mean field variables U and V:

$$U_{ia} = -\frac{1}{T} \frac{\partial E}{\partial V_{ia}} \qquad (1)$$

$$V_{ia} = \frac{e^{U_{ia}}}{\sum_b e^{U_{ib}}}. \qquad (2)$$

The MFA algorithm involves solving equations 1 and 2 at a series of progressively lower temperatures T: this process is known as temperature annealing. The critical temperature T_c, which sets the scale of T, is estimated by expanding equation 2 around the trivial fixed-point [13, 14] $V_{ia}^{(0)} = \frac{1}{N_a}$, where N_a is the number of possible states of each of the network neurons. For example, for the events defined by professor-class pairs (p, q) mapped onto classroom-timeslots (x, t), we have $N_p N_q$ neurons, each of which has $N_x N_t$ possible states, in which case $V_{pq;xt}^{(0)} = \frac{1}{N_x N_t}$.

Equations 1 and 2 can be solved iteratively using either synchronous or serial updating. The iterative dynamics to evolve the mean field variables toward a self-consistent solution is explained in detail by Peterson *et al.* [28]. The solutions correspond to stable states of the Hopfield network [17]. Observe from equation 2 that any solution to the MF equations respects a continuous version of the Potts condition

$$\sum_a V_{ia} = 1 \quad \forall\, i. \tag{3}$$

3.1 The Mean-Field Annealing Algorithm

The generic MFA algorithm appears in Figure 1. At high temperatures T, the mean-field solutions will be states near the fixed-point symmetrical maximum entropy state $V_{ia} = 1/N_a$. At low temperatures, finding a mean-field solution will be equivalent to using the Hopfield model, which is highly sensitive to the initial conditions and known to be ineffective for hard problems [17]. MFA improves over the Hopfield model by using annealing to slowly decrease the temperature in order to sidestep these problems.

These characteristics are similar to those of simulated annealing, which is no surprise since both it and the mean-field method compute thermal averages over Gibbs distributions of discrete states, the former stochastically and the latter through a deterministic approximation. It is therefore natural to couple the mean-field method with the concept of annealing from high to low temperatures.

In addition to the structure of the energy function, there are three major interdependent issues which arise in completely specifying a mean-field annealing algorithm for a timetabling problem:

- The values of the coefficients of terms in the energy function.
- The types of dynamics used to find solutions of the MFT equations at each T.
- The annealing schedule details, i.e. the initial temperature $T(0)$, the rules for deciding when to reduce T and by how much, and the termination criteria.

Peterson *et al.* [14] introduced a quantity called *saturation*, Σ, defined as

$$\Sigma = \frac{1}{N_i} \sum_{ia} V_{ia}^2 \,, \tag{4}$$

where N_i is the number of events (in this case the number of professor-class pairs). This characterizes the degree of clustering of events in time and/or space, $\Sigma_{min} = \frac{1}{N_a}$ corresponds to high temperature, whereas $\Sigma_{max} = 1$ means that all the V_{ia} have converged to 0 or 1 values, indicating that each event has been assigned to a space-time slot.

1. Choose a problem and encode the constraints into weights $\{w_{ij}\}$.
2. Find the approximate phase transition temperature by linearizing equation (2).
3. Add a self-coupling β-term if necessary. In a neural net, this corresponds to a feedback connection from a neuron to itself.
4. Initialize the neurons V_{ia} to high temperature values $\frac{1}{N_a}$ plus a small random term such as $rand[-1, 1] \times 0.001$; and set $T(0) = T_c$.
5. Until $(\Sigma \geq 0.99)$ do:
   - At each $T(n)$, update all U_{ia} and V_{ia} by iterating to a solution of the mean field equations.
   - $T(n+1) = \alpha T(n)$, we chose $\alpha = 0.9$
6. The discrete values S_{ia} that specify the schedule are obtained by rounding the mean field values V_{ia} to the nearest integer (0 or 1).
7. Perform *greedy heuristics* if needed to account for possible imbalances or rule violations.

Fig. 1. The Generic Mean-Field Annealing Algorithm

The first step of Figure 1 is to map the constraints of the problem into the neural net connection weights. In our implementation, at each T_n the MFA algorithm (Figure 1) performs one update per neural variable (defined as one sweep) with sequential updating using equations 1 and 2. After reaching a saturation value close to 1 (we chose $\Sigma = 0.99$) we check whether the obtained solutions are valid, i.e. $E_{hard} = 0$. If this is not the case the network is re-initialized and is allowed to resettle. We repeat this procedure a number of times until the best solution is found. A similar procedure was carried out on high school scheduling by Peterson *et al.* [14].

The MFA implementation was a little more complicated than the implementation of simulated annealing and the expert system, since it had many more parameters to handle, and it was often more difficult to find optimal values for these parameters. For example, one complication is the computation of the critical temperature T_c, which involved an iterative procedure of a linearized dynamic system. On the other hand, we observed that the convergence time was indeed much less than any of the convergence times of the simulated annealing using the three annealing schedules studied. For more details on our MFA implementation, see Ref. [10].

4 The Rule-Based System

We have implemented a fairly complex rule-based expert system for solving the timetabling problem, for three reasons. Firstly, it gives us a benchmark as to how well other methods do in comparison to this standard technique. Secondly, a simplified version of the rule-based system is used to provide sensible choices for moves in the simulated annealing algorithm, rather than choosing swaps completely at random, and this greatly improves the proportion of moves that are accepted. Thirdly, we have used this system as a preprocessor for simulated annealing, in order to provide a good initial solution.

Simulated annealing is a very time-consuming, computationally intensive procedure. Using an expert system as a preprocessor is a way of quickly providing a good starting point for the annealing algorithm, which reduces the time taken in the annealing procedure, and improves the quality of the result. Our results clearly support this rationale for the case of academic scheduling.

The rule-based expert system consists of a number of rules (or heuristics) and conventional recursion to assist in carrying out class assignments. We have developed this system specifically for the problem of academic scheduling. The basic data structures or components of the system are:

1. Distance matrix of values between each academic department and every other building under use for scheduling.
2. Class data structure of each class scheduled anywhere in campus. These structures are capable of linking with each other.
3. Room data structure of each room (regardless of type) involved in the scheduling process. Like classes, room structures are also linked with each other.
4. Data structures for time periods to keep track of which hour or time slot was occupied and which was not.
5. Department inclusion data structure giving department inclusion within other larger departments or colleges.
6. Students structures indicating classes of various degree of requirements and preferences for each student.

The basic function of the system is as follows: given data files of classes, rooms and buildings, department-to-building distance matrix, students data, and the inclusion data, using the abovementioned data structures, the system builds an internal database which in turn is used in carrying out the scheduling process. This process involves a number of essential sub-processes such as checking the distances between buildings, checking building, room type and hours occupied, checking and comparing time slots for any conflicts, checking rooms for any space conflict, and keeping track of and updating the hours already scheduled.

The rule-based system uses an iterative approach. The basic procedure for each iteration is as follows. The scheduling of classes is done by department, so each iteration consists of a loop over all departments. The departments are chosen in order of size, with those having the most classes being scheduled first.

The system first loops over all the currently unscheduled classes, and attempts to assign them to the first unoccupied room and timeslot that satisfies all the rules governing the constraints. Since constraints involving capacity of rooms are very difficult to satisfy, larger classes are scheduled first, to try to avoid not having large enough rooms later for those class sections with large enrollments.

In some cases the only rooms and timeslots that satisfy all the rules will already be occupied by previously scheduled classes. In that case, the system attempts to move one of these classes into a free room and timeslot, to allow the unscheduled class to be scheduled.

Next, the system searches through all the scheduled classes, and selects those that have a high cost, by checking the medium and soft constraints such as how closely the room size matches the class size, how many students have time conflicts, whether the class is in a preferred time period or a preferred building, and so on. Selecting threshold values for defining what is considered a "high" cost in each case is a subjective procedure, but it is straightforward to choose reasonable values. When a poorly scheduled class is identified, the system searches for a class to swap it with, so that the hard constraints are still satisfied, but the overall cost of the medium and soft constraints is reduced.

This process of swapping rooms continues provided all the rules are satisfied and no "cycling" (swapping of the same classes) occurs. Once all the departments have been considered, this completes one iteration. The system continues to follow this iterative procedure until a complete iteration produces no changes to the schedule.

There are many rules dealing with space and hours, type of room, and priority of room. Many are quite complex, but some of the basic rules, such as those implementing the hard constraints, can be quite straightforward – for example, the following is the basic rule for dealing with time and space conflicts for a room:

IF [room(capacity) > class(space-requested)] and [no time conflict in this room] THEN assign the room to the class.

When the rule-based system is used as a preprocessor, it produces a partial schedule as an output, since it is usually unable to assign all of the given classes to rooms and times slots. The output is divided into two parts: the first consists of classes, with their associated professors and students, assigned to various rooms; and the second is a list of classes that could not be assigned due to constraint conflicts.

5 Simulated Annealing

Simulated annealing (SA) has been widely used for tackling different combinatorial optimization problems, particularly academic scheduling [35, 7, 8]. The basic algorithm is described in Figure 2. The results obtained depend heavily on the cooling schedule used. We initially used the most commonly known and used schedule, which is the geometric cooling, but later tried adaptive cooling, as well as the method of geometric reheating based on cost [3].

A comprehensive discussion of the theoretical and practical details of SA is given in [1, 27, 32, 34]. It suffices here to say that the elementary operation in the Metropolis method for a combinatorial problem such as scheduling is the generation of some new candidate configuration, which is then automatically accepted if it lowers the cost (C), or accepted with probability $\exp(-\Delta C/T)$, where T is the temperature, if it would increase the cost by $\Delta(C)$. Also, in Figure 2, s is the current schedule and s' is a neighboring schedule obtained from the current neighborhood space $(\mathcal{N}_s)$ by swapping two classes in time and/or space.

Thus the technique is essentially a generalization of the local optimization strategy, where, at non-zero temperatures, thermal excitations can facilitate escape from local minima.

1. Generate an initial schedule s.
2. Set the initial best schedule $s^* = s$.
3. Compute cost of s : $C(s)$.
4. Compute initial temperature T_0.
5. Set the temperature $T = T_0$.
6. While *stop criterion* is not satisfied do:
   (a) Repeat *Markov chain length* (M) times:
      i. Select a random neighbor s' to the current schedule, $(s' \subset \mathcal{N}_s)$.
      ii. Set $\Delta(C) = C(s') - C(s)$.
      iii. If $(\Delta(C) \leq 0$ {downhill move}):
         – Set $s = s'$.
         – If $C(s) < C(s^*)$ then set $s^* = s$.
      iv. If $(\Delta(C) > 0$ {uphill move}):
         – Choose a random number r uniformly from $[0, 1]$.
         – If $r < e^{-\Delta(C)/T}$ then set $s = s'$.
   (b) Reduce (or update) temperature T.
7. Return the schedule s^*.

Fig. 2. The Simulated Annealing Algorithm

The SA algorithm has advantages and disadvantages compared to other global optimization techniques. Among its advantages are the relative ease of implementation, the applicability to almost any combinatorial optimization problem, the ability to provide reasonably good solutions for most problems (depending on the cooling schedule and update moves used), and the ease with which it can be combined with other heuristics, such as expert systems, forming quite useful hybrid methods for tackling a range of complex problems. SA is a robust technique, however, it does have some drawbacks. To obtain good results the update moves and the various tunable parameters used (such as the cooling rate) need to be carefully chosen, the runs often require a great deal of computer time, and many runs may be required.

Depending on the problem to which it is applied, SA appears competitive with many of the best heuristics, as shown in the work of Johnson *et al.* [21].

5.1 Timetabling Using the Annealing Algorithm

The most obvious mapping of the timetabling problem into the SA algorithm involves the following constructs:

1. a **state** is a timetable containing the following sets:
   - P: a set of professors.
   - C: a set of classes.
   - S: a set of students.
   - R: a set of classrooms.
   - I: a set of time intervals.
2. a **cost** or "energy" $E(P, C, S, R, I)$ such that:
   - $E(P)$: is the cost of assigning more than maximum number of allowed classes M_p to the same professor, plus scheduling one or more classes that cause a conflict in the professor's schedule.
   - $E(C)$: is the cost of scheduling certain classes at/within the same time period in violation of the exclusion constraint, for example.
   - $E(S)$: is the cost of having two or more classes conflict in time; plus cost of having in the schedule one or more classes that really don't meet the student's major, class requested, or class requirements; plus the cost of not having the classes evenly spread out over the week, etc.
   - $E(R)$: is the cost resulting from assigning room(s) of the wrong size and/or type to a certain class.
   - $E(I)$: is the cost of having more or less time periods than required, plus cost of an imbalanced class assignments (a certain period will have more classes assigned to than others, etc.).
3. A **swap** (or a move) is the exchange of one or more of the following: class c_i with class c_j in the set C with respect to time periods I_i and I_j, and/or with respect to classroom R_i and R_j, respectively. Generally, this step is referred to as class swapping.

Along with all of the necessary constraints, the simulated annealing algorithm also takes as input data the following: the preprocessor output in the form of lists of scheduled and non-scheduled classes and their associated professors and room types, a list of rooms provided by the registrar's office, a department to building distance matrix, a list of students and their class preferences, and a list of classes that are not allowed to be scheduled simultaneously.

To use simulated annealing effectively, it is crucial to use a good cooling schedule, and a good method for choosing new trial schedules, in order to efficiently sample the search space. We have experimented with both these areas, which are discussed in the following sections.

5.2 The Annealing Schedules

Three annealing schedules have been used in our experiments to update the temperature of the SA algorithm in Figure 2: geometric cooling, adaptive cooling, and adaptive reheating as a function of cost.

The first schedule we have used is **geometric cooling**, where the new temperature (T') of the SA algorithm is computed using

$$T' = \alpha T, \qquad (5)$$

where α $(0 < \alpha < 1)$ denotes the cooling factor. Typically the value of α is chosen in the range 0.90 to 0.99. This cooling schedule has the advantage of being well understood, having a solid theoretical foundation, and being the most widely used annealing schedule. Our results obtained from using this standard cooling schedule will be used as a baseline for comparison with those using the other two schedules, which allow the rate of cooling to be varied.

The second annealing schedule we used is the method of **reheating as a function of cost** (RFC), which was used for timetabling by Abramson et al. [3], but the ideas behind it are due to Kirkpatrick et al. [22, 23] and White [36]. Before introducing this schedule we first summarize a few relevant points on the concept of specific heat (C_H). Specific heat is a measure of the variance of the cost (or energy) values of states at a given temperature. The higher the variance, the longer it presumably takes to reach equilibrium, and so the longer one should spend at the temperature, or alternatively, the slower one should lower the temperature.

Generally, in combinatorial optimization problems, phase transitions [16, 26] can be observed as sub-parts of the problem are resolved. In some of the work dealing with the traveling salesman problem using annealing [24], the authors often observe that the resolution of the overall structure of the solution occurs at high temperatures, and at low temperatures the fine details of the solution are resolved. As reported in [3], applying a reheating type procedure, depending on the phase, would allow the algorithm to spend more time in the low temperature phases, thus reducing the total amount of time required to solve a given problem.

In order to calculate the temperature at which a phase transition occurs, it is necessary to compute the specific heat of the system. A phase transition occurs at a temperature $T(C_H^{max})$ when the specific heat is maximal (C_H^{max}), and this triggers the change in the state ordering. If the best solution found to date has a high energy or cost then the super-structure may require re-arrangement. This can be done by raising the temperature to a level which is higher than the phase transition temperature $T(C_H^{max})$. Generally, the higher the current best cost, the higher the temperature which is required to escape the local minimum. To compute the aforementioned maximum specific heat, we employ the following steps [3, 34, 27].

At each temperature T, the annealing algorithm generates a set of configurations $\mathcal{C}(T)$. Let C_i denote the cost of configuration i, $C(T)$ is the average cost at temperature T, and $\sigma(T)$ is the standard deviation of the cost at T.

At temperature T, the probability distribution for configurations is:

$$P_i(T) = \frac{e^{\frac{-C_i}{kT}}}{\sum_j e^{\frac{-C_j}{kT}}}. \tag{6}$$

The average cost is computed as:

$$< C(T) >= \sum_{i \in C} C_i P_i(T). \tag{7}$$

Therefore, the average square cost is:

$$< C^2(T) >= \sum_{i \in C} C_i^2 P_i(T). \tag{8}$$

The variance of the cost is:

$$\sigma^2(T) =< C^2(T) > - < C(T) >^2. \tag{9}$$

Now, the specific heat is defined as:

$$C_H(T) = \frac{\sigma^2(T)}{T^2}. \tag{10}$$

The temperature $T(C_H^{max})$ at which the maximum specific heat occurs, or at which the system undergoes a phase transition, can thus be found.

Reheating sets the new temperature to be

$$T = K \cdot C_b + T(C_H^{max}), \tag{11}$$

where K is a tunable parameter and C_b is the current best cost. Reheating is done when the temperature drops below the phase transition (the point of maximum specific heat) and there has been no decrease in cost for a specified number of iterations, i.e. the system gets stuck in a local minimum. Reheating increases the temperature above the phase transition (see equation 11), in order to produce enough of a change in the configuration to allow it to explore other minima when the temperature is reduced again.

The third cooling schedule we have tried is **adaptive cooling**. In this case, a new temperature is computed based on the specific heat, i.e. the standard deviation of all costs obtained at the current T. The idea here is to keep the system close to equilibrium, by cooling slower close to the phase transition, where the specific heat is large. There are many different ways of implementing this idea, we have chosen the approach taken by Huang et al. [18], which was shown to yield an efficient cooling schedule. Let T_j denote the current temperature, at step j of the annealing schedule. After calculating $\sigma(T_j)$ from equation 9, the new temperature T_{j+1} is computed as follows:

$$T_{j+1} = T_j \cdot e^{-\frac{aT_j}{\sigma(T_j)}}, \tag{12}$$

where a is a tunable parameter. Following suggestions by Otten and van Ginneken [27] and Diekmann *et al.* [6], $\sigma(T_j)$ is smoothed out in order to avoid any dependencies of the temperature decrement on large changes in the standard deviation σ. We used the following standard method to provide a smoothed standard deviation $\bar{\sigma}$:

$$\bar{\sigma}(T_{j+1}) = (1 - \omega)\sigma(T_{j+1}) + \omega\sigma(T_j)\frac{T_{j+1}}{T_j} \tag{13}$$

and set ω to 0.95. This smoothing function is used because it follows (from the form of the Boltzmann distribution, see [32, 36]) that it preserves the key relationship:

$$\frac{d}{dT}C(T) = \frac{\bar{\sigma}^2(T)}{T^2} = C_H \tag{14}$$

Note that reheating can be used in conjunction with any cooling schedule. We have used it with adaptive cooling.

5.3 The Choice of Moves

The performance of any application of simulated annealing is highly dependent on the method used to select a new trial configuration of the system for the Metropolis update. In order for the annealing algorithm to work well, it must be able to effectively sample the parameter space, which can only be done with efficient moves.

The simplest method for choosing a move is to swap the rooms or timeslots of two randomly selected classes. However this is extremely inefficient, since most of the time random swapping of classes will increase the overall cost, especially if we are already close to obtaining a valid solution (i.e. at low temperature), and will likely be rejected in the Metropolis procedure. This low acceptance of the moves means this simple method is very inefficient, since a lot of computation is required to compute the change in cost and do the Metropolis step, only to reject the move.

What is needed is a strategy for choosing moves that are more likely to be accepted. A simple example is in the choice of room. If we randomly choose a new room from the list of all rooms, it will most likely be rejected, since it may be too small for the class, or an auditorium when, for example, a laboratory is needed. One possibility is to create a subset of all the rooms which fulfill the hard constraints on the room for that particular class, such as the size and type of room. Now we just make a random selection for a room for that class only from this subset of feasible rooms, with an acceptance probability that is sure to be much higher. In addition, each class in our data set comes with a "type-of-space-needed" tag which is used along with other information to assign the class to the right room. This effectively separates the updates into independent sets based on room type, so for example, laboratories are scheduled separately from lectures. In our method we carry out the scheduling of lectures first, followed by scheduling of laboratories making sure that during the course of this process no lecture and its associated laboratory are scheduled in the same time period.

In effect, we have embedded a simple expert system into the annealing algorithm in order to improve the choice of moves, as well as using a more complex expert system as a preprocessor for the annealing step. When used to choose the moves for annealing, the main function of the rule-based system is to ensure that all the trial moves satisfy the hard constraints. Many of the rules dealing with the medium and soft constraints are softened or eliminated, since reducing the cost of these constraints is done using the Metropolis update in the annealing algorithm.

Another of the modifications to the rule-based system is that while the version used in the preprocessor is completely deterministic, the version used in choosing the moves for annealing selects at random from multiple possibilities that satisfy the rules equally well. This extra freedom in choosing new schedules, plus the extra degree of randomness inherent in the annealing update, helps prevent the system from getting trapped in a local minimum before it can reach a valid schedule, which is the problem with the standard deterministic rule-based system.

To improve further on the move strategy, we can take the subset of possible move choices that we have created for each class, and choose from them probabilistically rather than randomly. There may be certain kinds of moves that are more likely to be effective, so our move strategy is to select these moves with a higher probability. For example, swapping a higher level class (e.g. graduate) with a lower level class (e.g. a first or a second year type) generally has a higher acceptance, since there is little overlap between students taking these classes. Furthermore, we have experimented with two kinds of swaps, those that only involve classes offered by the same department or college and the second, swaps between classes of different departments and colleges.

Generally, the swap methods we have taken here can be considered as heuristics for pruning the neighborhood or narrowing the search space, which provides much more efficient moves and in turn an overall improvement in the results.

6 Experimental Results

Our computations were done with a number of goals in mind. The main objective was to provide a schedule which satisfied all hard constraints and minimized the cost of medium and soft constraints, using real-life data sets for a large university. We also aimed to find an acceptable set of annealing parameters and move strategies for general timetabling problems of this kind, and to study the effect of using a preprocessor to provide the annealing program with a good starting point. Finally, we wanted to make a comparison of the performance of the three different cooling schedules, geometric cooling, adaptive cooling, and reheating based on cost.

We spent quite some time finding optimal values for the various parameters for the annealing schedule, such as the initial temperature, the parameters controlling the rate of cooling (α for geometric cooling, a for adaptive cooling) and reheating (K), and the number of iterations at each temperature (for more details, see Ref. [11]). Johnson *et al.* [21] noted in their SA implementation for the traveling salesman problem (TSP) that the number of steps at each temperature (or the size of the Markov chain) needed to be at least proportional to the "neighborhood" size in order to maintain a high-quality result. From our experiments we found the same to be true for the scheduling problem, even though it is very different from the TSP. Furthermore, in a few tests for one semester we fixed the number of classes and professors but varied the number of rooms and time slots, and found that the final result improves as the number of iterations in the Markov chain becomes proportional to a combination of the number of classes, rooms and time slots. We also observed the same behavior when we fixed the number of rooms and time slots but varied number of classes.

Our study case involved real scheduling data covering three semesters at Syracuse University. The size and type of the three-semester data is shown in Table 1. Nine types of rooms were used: auditoriums, classrooms, computer clusters, conference rooms, seminar rooms, studios, laboratories, theaters, and unspecified types. Staff and teaching assistants are considered part of the set of professors. Third semester (summer) data was much smaller than other semesters, however, there were additional space and time constraints and fewer available rooms. Our data was quite large in comparison to data used by other researchers. For example, high school data used by Peterson and colleagues [13, 14] consists of approximately 1000 students, 20 different possible majors, and an overall periodic school schedule (over weeks). In the case of Abramson *et al.* [2], their data set was created randomly and was relatively small, and they stated that problems involving more than 300 tuples were very difficult to solve.

Table 1 lists all major components of the data we have used. Timetabling problems can be characterized by their *sparseness*. After the required number of lessons N_l have been scheduled, there will be $N_{sp} = (N_x N_t - N_l)$ spare space-time slots, hence, the sparseness ratio of the problem is defined as the ratio $N_{sp}/(N_x N_t)$. The denser the problem, the lower the sparseness ratio, and the harder the problem is to solve. Also, for dense problems, there is an additional correlation involving the problem size. Table 2 shows the sparseness of the three-semester data. For university scheduling, the sparseness ratio generally decreases as the data size (particularly the number of classes) increases, so the problem becomes harder to solve. Including student preferences makes the problem much harder, but these are viewed as medium constraints and thus are not necessarily satisfied in a valid solution.

Our overall results are shown in Tables 3 and 4. These tables show the percentage of classes that could be scheduled in accordance with the hard constraints. In each case (apart from the expert system, which is purely deterministic), we have done 10 runs (with the same parameters, just different random numbers), and the tables show the average of the 10 runs, as well as the best and worst results. The MFA results are different only due to having different initial conditions. Each simulated annealing run takes about 10 to 20 hours on a Unix workstation, while a single MFA run takes approximately an hour and an expert system run takes close to two hours.

Table 1. Size of the data set for each of the three semesters.

	First Semester	Second Semester	Third Semester
Rooms	509	509	120
Classes	3839	3590	687
Professors	1190	1200	334
Students	13653	13653	2600
Buildings	43	43	11
Schools and/or Colleges	20	21	17
Departments or Course Prefixes	143	141	108
Areas of Study (majors)	200	200	200

Table 2. The sparseness ratios of the problem for the data sets for each of the three semesters. Lower values indicate a harder problem.

Academic Time Period	Sparseness ratio
First Semester	0.50
Second Semester	0.53
Third Semester	0.62

Table 3. Percentage of classes scheduled using the different methods. The averages and highest and lowest values were obtained using 10 independent runs for simulated annealing (SA) and mean-field annealing (MFA). The expert system (ES) is deterministic so the results are from a single run. *No preprocessor was used with the three methods.*

Academic Time Period	Algorithm	Scheduled (average) %	Highest Scheduled %	Lowest Scheduled %
First Semester	SA (geometric)	65.00	67.50	56.80
	SA (adaptive)	67.80	70.15	61.20
	SA (cost-based)	70.20	72.28	68.80
	ES	76.65	76.65	76.65
	MFA	65.60	71.00	61.00
Second Semester	SA (geometric)	65.65	68.00	57.10
	SA (adaptive)	68.50	70.10	60.77
	SA (cost-based)	75.14	77.68	70.82
	ES	79.00	79.00	79.00
	MFA	67.20	75.00	65.00
Third Semester	SA (geometric)	83.10	86.44	68.50
	SA (adaptive)	85.80	89.00	70.75
	SA (cost-based)	91.20	95.18	85.00
	ES	96.80	96.80	96.80
	MFA	88.00	95.00	82.00

Table 4. Percentage of scheduled classes, averaged over 10 runs of the same initial temperature and other parameters, for three terms using simulated annealing *with an expert system as preprocessor.*

Academic Time Period	Algorithm	Scheduled (average) %	Highest Scheduled %	Lowest Scheduled %
First Semester	SA (geometric)	93.90	95.12	85.20
	SA (adaptive)	98.80	99.20	95.00
	SA (cost-based)	100.0	100.0	100.0
Second Semester	SA (geometric)	95.00	98.95	89.40
	SA (adaptive)	99.00	99.50	98.50
	SA (cost-based)	100.0	100.0	100.0
Third Semester	SA (geometric)	97.60	98.88	90.90
	SA (adaptive)	100.0	100.0	100.0
	SA (cost-based)	100.0	100.0	100.0

As expected, each of the methods did much better for the third (summer) semester data, which has a higher sparseness ratio. Our results also confirm what we expected for the different cooling schedules for simulated annealing, in that adaptive cooling performs better than geometric cooling, and reheating improves the result even further.

When a random initial configuration is used, simulated annealing performs very poorly, even worse than the expert system (ES). However, there is a dramatic improvement in performance when a preprocessor is used to provide a good starting point for the annealing. In that case, using the best cooling schedule of adaptive cooling with reheating as a function of cost, we are able to find a valid class schedule every time.

In the case of mean-field annealing, the overall results are generally below those of SA and ES. In addition, we have found in the implementation of this method that the results were quite sensitive to the size of the data as well the type of constraints involved. If we confine ourselves to the set of hard constraints, the results are as good as or even better than the other methods. However if we take into account the medium and soft constraints, that is, the overall cost function, this method does not perform as well.

Student preferences are included only as medium constraints in our implementation, meaning that these do not have to be satisfied for a valid solution, but they have a high priority. For the valid schedules we have produced, approximately 75% of the student preferences were satisfied. This is reasonably good (particularly since other approaches do not deal with student preferences at all), but we are working to improve upon this result.

7 Conclusions

We have successfully applied simulated annealing to the difficult problem of academic scheduling for a large university. Feasible schedules were obtained for real data sets, including student preferences, without requiring enormous computational effort.

Mean-field annealing works well for small scheduling problems, but does not appear to scale well to large problems with many complex constraints. For this problem, both simulated annealing and the rule-based system were more effective than MFA. It is more difficult to tune the parameters for MFA than for simulated annealing, and because of the complexity and size of the Potts neural encoding, there seems to be no clear way of preserving the state of a good initial configuration provided by a preprocessor when using MFA.

Using a preprocessor to provide a good initial state greatly improved the quality of the results for simulated annealing. In theory, using a good initial state should not be necessary, and any initial state should give a good result, however in practice, we do not have an ideal cooling schedule for annealing, or an ideal method for choosing trial moves and efficiently exploring the search space, and there are restrictions on how long the simulation can take. In general, for very hard problems with large parameter spaces that can be difficult to search efficiently, and for which very slow cooling would be much too time-consuming, we might expect that a good initial solution would be helpful. We used a fairly complex rule-based expert system for the preprocessor, however the type of preprocessor may not be crucial. Other fast heuristics could possibly be used, for example a graph coloring approach [25], or it may be possible to just utilize the schedule from the same semester for the previous year. A modified version of the rule-based system was used to choose the trial moves for the simulated annealing, and the high acceptance rate provided by this system was crucial to obtaining good results.

As expected, for the simulated annealing, adaptive cooling performed better than geometric cooling, and using reheating improved the results even further. The best results were obtained using simulated annealing with adaptive cooling and reheating as a function of cost, and with a rule-based preprocessor to provide a good initial solution. Using this method, and with careful selection of parameters and update steps, we were able to generate solutions to the class scheduling problem using real data for a large university. None of the other methods were able to provide a complete solution.

Our main conclusion from this work is that simulated annealing, with a good cooling schedule, optimized parameters, carefully selected update moves, and a good initial solution provided by a preprocessor, can be used to solve the academic scheduling problem at a large university, including student preferences. Similar approaches should prove fruitful for other difficult scheduling problems.

Acknowledgments

The first author is very grateful for the valuable discussion and help of Robert Irwin in converting and formatting the registration data prior to the scheduling process. We also would like to thank Andrew Gee and Martin Simmen for the useful comments and suggestions, and Carsten Peterson for the pointers and comments about his papers. Many thanks go to Karen Bedard for providing us with the data and answering so many questions we had about it, Meg Cortese for providing us with a set of building constraints for various departments, and Prof. Ben Ware, Vice President for Research and Computing at Syracuse University, for his support and encouragement.

References

1. Aarts, E. H., J. Korst, and P. J. van Laarhoven, "Simulated annealing," in *Local Search in Combinatorial Optimization*, E. H. Aarts and J. K. Lenstra (eds.), John Wiley and Sons, 1997.
2. Abramson, D., "Constructing school timetables using simulated annealing: sequential and parallel algorithms," *Management Science* 37(1), 98-113, 1991.
3. Abramson, D., H. Dang, and M. Krishnamoorthy, "An Empirical Study of Simulated Annealing Cooling Schedules," Griffith Univ. report, Nathan, Qld, Aus. 1994; "Simulated Annealing Cooling Schedules for the School Timetabling Problem," submitted to Asia Pacific Journal of Operations Research, 1996.
4. Burke, E., and P. Ross, eds., *Practice and Theory of Automated Timetabling, First International Conference, Edinburgh, 1995 : Selected Papers*, Lecture Notes in Computer Science no. 1153, Springer, New York, 1996.
5. de Werra, D., "An introduction to timetabling," *European Journal of Operational Research* 19, 151-162, 1985.
6. Diekmann, R., R. Lüling, and J. Simon, "Problem independent distributed simulated annealing and its applications," in *Applied Simulated Annealing*, R. V. Vidal ed., Lecture Notes in Economics and Mathematical Systems, no. 396, Springer-Verlag, 1993.
7. Dowsland, K., "Using Simulated Annealing for Efficient Allocation of Students to Practical Classes", Working Paper, Statistics and OR Group, European Business Management School, University College of Swansea, UK, 1994.
8. Dowsland, K. and J. Thompson, "Variants of Simulated Annealing for the Examination Timetabling Problem," Working Paper, Statistics and OR Group, European Business Management School, University College of Swansea, UK, 1994.
9. Eiselt H. A., and G. Laporte, "Combinatorial Optimization Problems with Soft and Hard Requirements," *J. Operational Research Society*, vol. 38, No. 9, pp. 785-795, 1987.
10. Elmohamed, S., G. C. Fox, P. Coddington, "Course Scheduling using Mean-Field Annealing, Part I: algorithm and Part II: implementation," Northeast Parallel Architectures Center technical report SCCS-782, Syracuse University, Syracuse, NY, 1996.
11. Elmohamed, S., P. Coddington, G.C. Fox, "Academic Scheduling using Simulated Annealing with a Rule-Based Preprocessor", Northeast Parallel Architectures Center technical report SCCS-781, Syracuse University, Syracuse, NY, 1997.

12. Gee, Andrew, private communication.
13. Gislén, L., B. Söderberg, C. Peterson, "Teachers and Classes with Neural Nets," *International Journal of Neural Systems* 1, 167 (1989).
14. Gislén, L., B. Söderberg, C. Peterson, "Complex scheduling with Potts neural networks," *Neural Computation*, 4, 805-831, 1992.
15. Hertz, J., A. Krogh and R. Palmer, *Introduction to the Theory of Neural Computation*, Addison-Wesley, Redwood City, CA, 1991.
16. Hogg, T., B. Huberman, and C. Williams (editors), Artificial Intelligence, special issue on *Phase transitions and the search space*, p. 81, 1996.
17. Hopfield, J. J., and D. W. Tank, "Neural Computation of Decisions in Optimization Problems," *Biological Cybernetics* 52, 141 (1985).
18. Huang, M., F. Romeo, and A. Sangiovanni-Vincentelli, "An efficient general cooling schedule for simulated annealing," *Proc. of the IEEE International Conference on Computer Aided Design (ICCAD)*, pp. 381-384, 1986.
19. Johnson, D., C. Aragon, L. McGeoch, and C. Schevon, "Optimization by Simulated Annealing: an Experimental Evaluation, Part I (Graph Partitioning)," *Operations Research* 37, 865-892 (1989).
20. Johnson, D., C. Aragon, L. McGeoch, and C. Schevon, "Optimization by Simulated Annealing: an Experimental Evaluation, Part II (Graph Coloring and number partitioning)," *Operations Research* 39, No. 3, 865-892 (1991).
21. Johnson, D., and L. McGeoch, "The Traveling Salesman Problem: A Case Study in Local Optimization," in *Local Search in Combinatorial Optimization*, E. H. Aarts and J. K. Lenstra (eds.), Wiley and Sons, 1997.
22. Kirkpatrick, S., C. D. Gelatt, Jr., and M. P. Vecchi, "Optimization by Simulated Annealing," *Science* 220, 671-680, (13 May 1983).
23. Kirkpatrick, S., "Optimization by simulated annealing: Quantitative studies," *J. Stat. Physics* 34, 976-986 (1984).
24. Lister, R.,"Annealing Networks and Fractal Landscapes," *Proc. IEEE International Conference on Neural Nets*, March 1993, Vol. I, pp 257-262.
25. Miner, S., S. Elmohamed, and H. W. Yau, "Optimizing Timetabling Solutions Using Graph Coloring," 1995 NPAC REU program, NPAC, Syracuse University, Syracuse, NY, 1995.
26. Mouritsen, O. G., *Computer Studies of Phase Transitions and Critical Phenomena*, Springer-Verlag, Berlin, 1984.
27. Otten, R., and L. van Ginneken, *The Annealing Algorithm*, Kluwer Academic Publishers, 1989.
28. Peterson, C., and B. Söderberg, "Artificial Neural Networks and Combinatorial Optimization Problems," *Local Search in Combinatorial Optimization*, E.H.L. Aarts and J.K. Lenstra (eds.), Wiley and Sons, 1997.
29. Peterson, C., and B. Söderberg, "A New Method for Mapping Optimization Problems onto Neural Nets", *International Journal of Neural Systems* 1, 3 (1989).
30. Schaerf, A., "A survey of automated timetabling," Department of Software Technology, Report CS-R9567, CWI, Amsterdam, The Netherlands.
31. Simmen, Martin, Personal Communication.
32. Sorkin, G., Theory and Practice of Simulated Annealing on Special Energy Landscapes, PhD. Thesis, Dept. of Electrical Engineering and Computer Science, University of California, Berkeley, July 1991.
33. Thompson, J., and K. Dowsland, "General Cooling Schedules for Simulated Annealing Based Timetabling Systems," *Proceedings of the 1st International Conf. on the Practice and Theory of Automated Timetabling*, Napier Univ., Edinburgh 1995.

34. van Laarhoven, P. J. and E. H. Aarts, Simulated Annealing: Theory and Applications. D. Reidel, Dordrecht, 1987.
35. Vidal, R. V. ed., *Applied Simulated Annealing*, Lecture Notes in Economics and Mathematical Systems no. 396, Springer-Verlag, 1993.
36. White, S. R., "Concepts of scale in simulated annealing," *Proceedings of the IEEE International Conference on Circuit Design*, pp 646-651, 1984.

Evolutionary Computation
(Population Based Methods)

Some Observations about GA-Based Exam Timetabling

Peter Ross[1], Emma Hart[1]*, and Dave Corne[2]

[1] Department of AI, University of Edinburgh, 5 Forrest Hill, Edinburgh, EH1 2QL
{peter,emmah}@dai.ed.ac.uk
[2] Dept. of Computer Science, University of Reading, Whiteknights
Reading, RG6 6AY
D.W.Corne@reading.ac.uk

Abstract. Although many people have tried using genetic algorithms (GAs) for exam timetabling, far fewer have done systematic investigations to try to determine whether a GA is a good choice of method or not. We have extensively studied GAs that use one particular kind of direct encoding for exam timetabling. Perhaps not surprisingly, it emerges that this approach is not very good, but it is instructive to see why. In the course of this investigation we discovered a class of solvable problems with interesting properties: our GAs would sometimes fail to solve some of the moderately-constrained problems, but could solve all of the lightly-constrained ones and all of the highly-constrained ones. This is despite the fact that they form a hierarchy: those erratically-solved problems are subproblems of the easily-solved but highly-constrained ones. Moreover, some other non-evolutionary approaches also failed on precisely the same sets. This, together with some observations about much simpler graph-colouring methods based on the Brelaz algorithm, suggest some future directions for GA-based methods.

1 Introduction

Setting up a genetic algorithm to solve some OR problem can be very difficult. A representation must be devised, genetic operators and parameters must be chosen, and there are even many varieties of genetic algorithm to consider: steady-state, generational, breeder, island, cellular, CHC [14], and so on. Many authors have tried using GAs to solve various kinds of timetabling problem – see [4] for several examples. However, few authors have yet done any systematic study to see how reliable a GA might be at solving some class of timetabling problems or even whether it is easy to configure a GA to solve a particular problem. And Dowsland [12] rightly points out that GA researchers often fail to make proper comparisons with alternative methods, though some examples do exist, for example [6].

This paper outlines the history of an extensive investigation of GAs that use a particularly simple and direct representation. Discovering the limitations of

---

* formerly Emma Collingwood

the approach is instructive. Some of the weaknesses of the representation are exposed, but it would also seem that sometimes an easy-looking problem can be deceptively difficult for a variety of algorithms, not just GAs.

This paper focuses on exam timetabling: events must not clash, there may be seating limitations to obey, and it may also be desirable to try to minimise some measure of spread such as the number of students who have to take consecutive exams in any day. If a student takes two exams in consecutive slots on the same day, we call it a *near-clash*.

2 The representation

We discuss a straightforward, direct representation. For examples of other representations used in GA timetabling, the reader is referred again to [4]; for instance, [1] describe an indirect representation using a permutation of lectures and suggestion lists as to where to place them, and [3, 13] both describe more complex direct representations. In the representation discussed here, a timetable of E exams is represented as an array of E numbers. The i-th number indicates which timeslot the i-th event is to be placed in. For GA purposes such a representation can be evaluated by simple penalty-function methods, and we chose to do this rather than adopting any more contentious or exotic means such as 'repairing' the array to force some constraints to hold. Penalty functions are regarded with suspicion in the GA world, even though they are very respectable in the simulated annealing community. GA folklore also suggests that it is hard to choose penalty values, but our experience has been that it is a matter of common sense. For example, if a clash earns a penalty of 1 then it doesn't make sense to assign a penalty of 0.1 for some near-clash. That would mean that fixing eleven near-clashes would appear to be more worthwhile than fixing one clash, and a little thought suggests that in real problems there are often many moves that can fix substantial numbers of near-clashes at a stroke. A near-clash penalty of 0.01 makes more practical sense. And penalty function methods are at least computationally cheap.

3 Some history

With the aid of grant from the UK EPSRC we set about exploring GAs that used such a representation for exam timetabling. We did explore issues of representation [11, 19] early on, considering GAs which varied the number of timeslots against those which varied the number within allowable limits on the number of constraint violations and against those given a fixed number of timeslots. This last group performed well on some modestly-sized local problems, and we found that it was not too sensitive to the choices of penalties, provided that a form of directed mutation was used instead of random mutation [10, 9, 23, 22], ie a memetic algorithm. In particular, a gentle pressure to improve quality was by far the best; local optimisation caused deadlocks that drift would spread through the population (as yet unpublished hand analysis). We then studied performance on a range of benchmark problems modelled on the Department of AI's own timetabling problems with 50-100 events and ca. 200 students and 36 time slots; approaches included varying the number of events, students and timeslots independently. This produced a semi-linear performance graph, enabling us to predict performance on much larger problems. Tests with Edinburgh University data (ca. 1000 events, 9000 students whose exams were not already fixed by convention) showed that the predictions were reasonable. Further tests with such large problems contributed by other universities were also very encouraging, with solutions in an hour or so on a modest Sparc workstation.

An MSc project [17] showed that, for the case of problems involving only binary constraints (ie pairs of events must not clash), conventional graph-colouring methods would outperform a GA easily, even if a GA was helped by finding cliques in the graph and keeping them close in the chromosomes. Pre-colouring those cliques did help the GA greatly, but for large problems the cost of finding them is too great. The GA is best handling problems that include non-binary constraints too; graph-colouring methods do not extend particularly easily to these.

A GA seemed to be a good option for handling large timetabling problems that involved significant numbers of non-binary constraints as well as binary ones and seemed to perform well on our real-world benchmarks [10, 9, 11]. But what about those problems that involved only binary constraints? We showed that a stochastic hillclimber would often outperform a GA [8, 20] but that a GA was a better choice if multiple solutions were wanted (see also [27]). Further investigation showed first that there were small classes of problem on which a GA would outperform a stochastic hillclimber or simulated annealing [7], and that each choice of algorithm could be significantly improved by using an initialisation strategy that was neither random nor greedy but tried just a little to improve on random candidates ('peckish strategies', [8]). We then showed that, for parametrically-generated timetabling problems, there were regions of the parameter space where a GA would outperform other methods – typically regions of very high constrainedness [21].

In the course of this work we explored a wide variety of GA choices. What often seemed to work well was a steady-state GA with a fairly small population, perhaps 20, using tournament selection of size 2 (equivalent to linear ranking), with two-point, uniform or no crossover producing one child which is then subjected probabilistically to a directed mutation as mentioned above. The mutation step would choose a single event and then find a slot to which to move it. We considered various methods. Choosing an event could be done at random, or by considering the penalty associated with each event and using either roulette-wheel selection or tournament selection of size 2 to choose one. Finding a slot could be done at random, or by using tournament selection to find a better slot for it, or by extensive searching to find an ideal slot for it.

It appeared that this kind of GA based on using the direct representation was at best a mediocre choice of algorithm for handling problems that involved only binary constraints, that is, graph-colouring problems. But it seemed worth investigating a bit further to clarify why. To do so we constructed some classes of artificial problems, two of which were very revealing.

4 The 'pyramidal chain' class

Consider graph-colouring problems of the following kind. There are C cliques in all, each of size N. Cliques 1 and 2 overlap by $(N - 1)$ nodes; cliques 2 and 3 overlap by 1 node; cliques 3 and 4 overlap by $(N - 1)$ nodes; and so on, until finally cliques C and 1 overlap by 1 node (which means that C must be even and at least 4). Figure 1 illustrates the idea for $N = 5$.

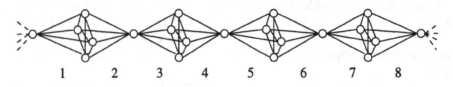

1   2   3   4   5   6   7   8

Fig. 1. A 'pyramidal chain' problem

It should be clear that a simple method such as the Brelaz algorithm [2] will solve such problems without trouble, using N colours. However, the kind of GA outlined above could not solve even modestly-sized examples of this problem, for instance with $C = 20, N = 6$. It is also fairly clear why that should be so. Consider those nodes where two adjacent cliques overlap by a single node. All such nodes must have the same colour in any solution. To see this, observe that the colour of the node shared between cliques 2 and 3 is fully determined by the other $(N-1)$ nodes in clique 3. These also determine the colour of the remaining node in clique 4, which is the single node shared by cliques 4 and 5.

However, nothing is pushing the GA towards ensuring that all these single nodes have the same colour. Each solved clique increases fitness appreciably, and so the GA spends its efforts on solving the individual cliques *in parallel*, effectively ignoring the need to co-ordinate them until it is too late. By that time the population has moved too far towards uniformity; co-ordination of the linking nodes can only be achieved by recolouring whole cliques, and the kind of mutation which changes a single colour at a time cannot hope to do that effectively. With hindsight, or *a priori* knowledge of the problem structure, it would be perhaps be possible to design a more suitable representation or invent specific operators that could "push" a GA towards a solution in *this* particular problem instance; however, in the general case, we do not know what a problem solution should look like, and it would be more productive to find a GA that could find solutions to a *range* of problems with differing characteristics.

A possible way to discourage the GA from getting into this hole is to use a more ambitious form of mutation. It needs to be able to change several colours at a time; it needs to focus more on changing the colours of wrongly-coloured nodes; and it needs to be applied more strongly as the population approaches convergence. The following *cataclysmic adaptive* mutation scheme satisfies these criteria. When the child of two parents is to be mutated, an effective mutation rate $\alpha = (1 - d/L)$ is calculated, where L is the length of the chromosome and d is the number of gene positions at which the two parents differ. Thus the rate is 1.0 if the parents are identical and 0.0 is they are maximally different. Then, for each i, if P_i is the penalty caused by gene (ie node, exam) i in the child, that gene is mutated to a better value (using one of the schemes mentioned earlier) with a probability $\alpha(0.1 + P_i)/(\sum P_i)$.

It turns out that this can indeed solve problems such as the $C = 20, N = 6$ case. However, it should be obvious that it will fail if C is made large enough; and this is despite the fact that as C increases the edge density tends to 0. This class of problems clearly exposes the weakness of the direct representation.

5 The 'sequence of cliques' class

The second class produces very much more surprising results. It is constructed in the following way. First, C non-overlapping cliques each of size N are created. Then edges are created probabilistically between every pair of cliques, in such a way as to ensure that the resulting problem is solvable with N colours. If the nodes of each clique are numbered $1 \cdots N$ then no edge is ever created between node i in one clique and node i in another. This means that a permutation which solves one clique also solves all the others and doesn't violate any inserted inter-clique edge. So, between two given cliques there are C forbidden edges and $C(C-1)$ permissible edges. Each of the permissible edges is created with a fixed probability p. The expected number of edges is therefore

$$\frac{NC(C-1)}{2} + \frac{pN(N-1)C(C-1)}{2}$$

which, for $p = 1.0$, produces a maximum edge density of $(NC - N)/(NC - 1)$.

The edge-insertion process essentially assigns a random number $0 \le r_e \le 1$ to each permissible edge e, and that edge is then included if $r_e < p$. The sequence of random numbers used is independent of p. This means that if we generate, say, 10 problems for $p = p_1$ and 10 problems for $p = p_2 > p_1$, then the problems for p_2 *each contain the corresponding problem for* p_1.

Consider, therefore, an example in which we generate 10 problems for each $p = 0.00, 0.05, 0.10, 0.15, \cdots, 0.95, 1.00$. Figure2 shows how one particular GA performed. The horizontal axis is the probability. The vertical axis is the number of clashes remaining at convergence. The error bars indicate the range of performance over the 10 problems for each p and the diamonds, joined by the dotted line, indicate the average of the indicated measure over the 10 problems.

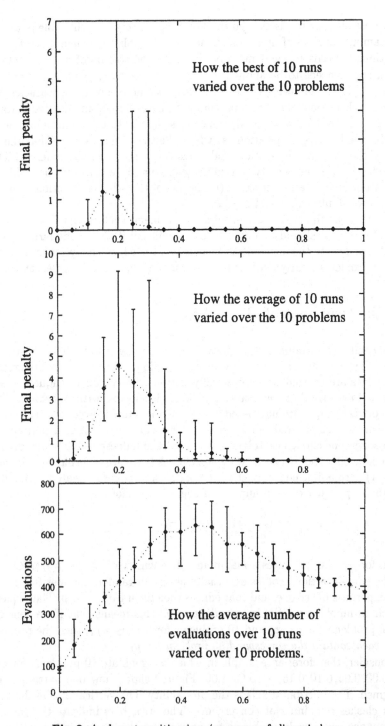

Fig. 2. A phase-transition in a 'sequence of cliques' class

As this shows, the GA in question fails to solve some of the problems for a certain range of p. This is despite the fact that the GA can (easily) solve all the problems for the higher values of p. That is, the GA can solve all the problems for (say) $p = 0.80$ but *fails on certain sub-problems of these!*. Even more surprisingly:

- *all the GAs mentioned above show similar performance graphs, even when using cataclysmic adaptive mutation. The size of the peak and the breadth of the fallible region can vary, but the region is always around $p = 0.20$. The location of the fallible region depends on N and C but is in the lower part of the spectrum of probabilities. Performance is similar but worse if crossover is turned off altogether.*
- *The Brelaz algorithm also fails, just on certain of the problems in the fallible region.*
- *Carter's non-evolutionary 'knockback' algorithm, described in [5], also fails on some of the problems – but only in the same fallible region.*
- *An approach suggested by Michael Trick, which finds maximal cliques and colours them first, but backtracks as necessary, finds those problems in the fallible region much harder than the others. Because it can backtrack exhaustively, and because the problems are all solvable, it does manage to solve them; but the awkward problems in the fallible region require far more backtracking than any of the others.*

As an illustration, Figure 3 shows the performance of Carter's and Trick's algorithms on the same problems used in Figure 2.

This is very intriguing. It suggests that all these algorithms share the same sort of weakness – perhaps they make some early, innocuous-looking but wrong decisions that lead them into trouble. But the fact that they can all solve the highly-constrained problems easily, yet fail on certain sub-problems of those that they can solve, is very curious and deserves further study. It also suggests that it may not always be a GA's fault (so to speak) if it fails to solve a problem. For instance, the occurrence of similar phase-transition regions in other classes of binary constraint satisfaction problems has been widely reported, for example see [16, 18, 25], in a Special Issue of the Journal of Artificial Intelligence on phase transitions in combinatorial optimization. In many of the problems discussed in this issue, a phase transition in the difficulty of solving problems is observed, in which the transition is found when moving from a region in which all problems have many solutions to a region in which almost all problems have no solutions.

6 Carter's problem set

Although the 'sequence of cliques' class is very interesting, there remains the question of how any of the GAs that use the simple direct representation perform on other real-world problems. It was possible to get good results acceptably quickly on Edinburgh University's problem that had just over 1000 exams and around 9000 students, but perhaps this problem is unrepresentative

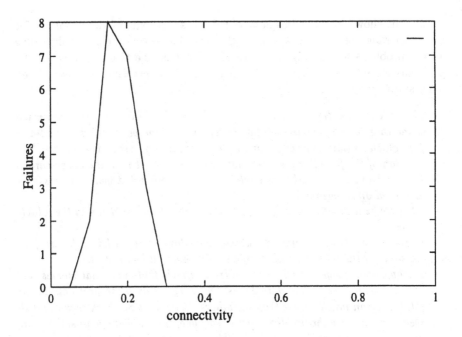

M.Carter Algorithm

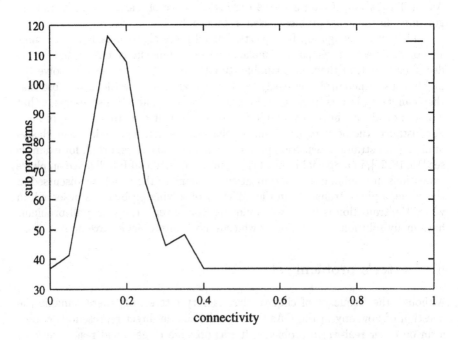

Trick Algorithm

Fig. 3. Performance of Carter's and Trick's algorithms

in some way. A set of real-world exam timetabling problems is available from
ftp://ie.utoronto.ca/mwc/testprob/, by courtesy of Mike Carter. An earlier
version of this collection, dating from around 1994, exists in some places. The
earlier version appeared to have flawed versions of the EAR, STA and YOR
problems and a somewhat different version of the UTAS problem. The naming
also changed, but in a recognisable way.

Some results on these problems have been published – see for example [5, 3,
26]. Table 1 gives some information about the current versions of these problems. The column headed 'Edges' gives the number of edges in the graph of the
problem, where each exam is a vertex, each edge represents the fact that those
two vertexes cannot be in the same timeslot, and edges are not duplicated.

Problem	Exams	Students	Edges	Slots	Seats	Brelaz Colours	Other GC
car-f-92	543	18419	20305	40	2000	30	28
car-s-91	682	16926	29814	51	1550	31	28
ear-f-83	190	1125	4793	?	?	23	22
hec-s-92	81	2823	1363	18	?	19	17
kfu-s-93	461	5349	5893	20	1955	19	19
lse-f-91	381	2726	4531	?	?	19	17
pur-s-93	2419	30032	86261	?	5000	35	35
rye-s-93	486	11483	8872	?	?	22	21
sta-f-83	139	611	1381	?	?	13	13
tre-s-92	261	4362	6131	35	655	23	20
uta-s-92	622	21266	24249	38	2800	31	32
ute-s-92	184	2750	1430	?	?	10	10
yor-f-83	181	941	4706	?	?	20	19

Table 1. Carter's real-life exam timetable problems

For these problems the following conditions are sometimes assumed, even
though these are not always what the original problem required [5]:

– there are three timeslots per day;
– in any one day, a student should not have to take two consecutive exams
  if at all possible. Each instance of such a near-clash counts as a separate
  violation of this soft constraint.
– there is a limit on the total number of seats available in any slot. This figure
  is given in the column headed 'Seats'.

We compare some of the reported results in the remainder of this section,
and also contrast the results with those by a modified graph colouring algorithm.

Results Reported by Carter Carter et al [5] use different soft and hard constraints from these, and give comparative results for various non-evolutionary
algorithms derived from graph-colouring heuristics. They tried those algorithms

with and without backtracking, and both as simple graph-colouring problems and as weight-minimisation problems in which weights were attached to any instance of a student having to write two exams 5 or less slots apart. Their results do not say how small the final weighted sums were, however; they just give the number of slots found necessary by the algorithm, so comparisons need to be done with caution.

Among many other points they observe that backtracking helps and that simple non-backtracking graph-colouring can often fail to find the smallest number of slots. For example the Brelaz algorithm [2] merely colours the graph without backtracking, choosing vertexes first by colour degree (alias saturation degree; that is, the number of different colours already present among its neighbours) with tie-breaking by vertex degree (the number of distinct edges from that vertex). The number of slots this produces, if slot capacity constraints are ignored and backtracking is forbidden, is given in Table 1 in the column headed 'Brelaz Colours'. The column headed 'Other GC' reports the smallest number of slots required by the graph-colouring algorithms reported in [5], with backtracking permitted. One interesting point to note is that the uta-s-92 problem needs only 31 slots according to our Brelaz program, even taking the seating limitation into account.

Results Reported by Burke The column headed 'Slots' shows the number of slots used in Burke et al [3], when the restriction on seats per slot is enforced and the aim is to minimise the number of near-clashes as well as avoiding all clashes. For example, Burke et al report that their memetic GA can solve the tre-s-92 problem using 35 slots, finding a solution with only 3 near-clashes using 4500 evaluations (this is one of the easiest problems in the set).

Direct Representation It turns out that the GAs using the direct representation do not perform particularly well on these problems, even when the seating capacity constraint is ignored. Even with the cataclysmic adaptive mutation it can take an hour or more on a 166Mhz PC to eliminate all clashes and reduce the number of near-clashes appreciably, and even then the results are typically not as good as those cited in [3]. It would seem that the problem is again a failure to co-ordinate the different parts of a solution until it is too late, leaving mutation to do the final work in an unreasonably long time.

Modified Brelaz Algorithm Some experiments with Brelaz-like algorithms that obey the seat capacity constraints produce suggestive results not too far removed from Burke et al's results, and very quickly indeed. Consider the following algorithm:

- assign exams to slots according to colour (saturation) degree. Ties are broken according to the number of instances of arcs at a node. This is not the same as the number of students taking the exam. If there are 20 students taking an exam and each is taking the same three other exams there will be only three arcs, but there will be 60 instances of arcs.

- slots are not chosen consecutively. In order to try to reduce near-clashes, whenever a suitable slot is being sought for an exam, the algorithm begins by considering the first slot of each day. If none of those are suitable, because of clashes or seating limitations, then it examines the last slot of each day. Only if that fails does it finally examine the middle slot of each day.

Both this and the more conventional Brelaz algorithm (in which ties are broken by node degree, slots are merely examined consecutively but the seating limitations are enforced) solve the tre-s-92 problem quickly.

The conventional Brelaz algorithm produces an answer using 27 slots, involving 1355 near-clashes, and in which 16 of the slots are at full capacity. The modified Brelaz algorithm produces an answer using 34 slots, involving just 87 near-clashes. But 34 slots requires 12 days. If this same algorithm is allowed to use the other two slots on the last day then the slot-choosing process leads to a different result in which all 12 days are still used, there are just 24 near-clashes .. and the middle slot of every day except the first two have no exams in them! The exams in the middle slots of the first two days produce all the near-clashes. It is somewhat surprising that weakening the requirements a little can lead to what is arguably a better result. The algorithm is also fast, taking just two seconds on a 166Mhz (non-MMX) Pentium PC.

The tre-s-92 problem is however undemanding; there are many exams that involve comparatively few students, so it is not difficult to meet the seating capacity constraints. Consider a rather more demanding one, kfu-s-93. The conventional Brelaz algorithm solves this problem using 19 slots, with 7300 near-clashes (in four seconds) and one slot at full capacity. The modified algorithm fails to solve it, even with 21 slots. The reason would seem to be the difficulty of packing exams so as to honour the seating capacity. The problem contains a number of reasonably large exams, and the algorithm spreads them over slots in such a way that there comes a point at which no feasible slot has enough seating capacity left to handle another large exam. What would seem to be needed is a variant algorithm which pays more attention to the implicit bin-packing problem early on. It is worth noting, however, that the modified algorithm can solve it if allowed 23 slots; this produces an answer involving 1224 near-clashes, with 6 slots at full capacity; and the very last slot contains a single exam, which it turns out can be moved to any of three other slots without clashes, adding at most 31 further near-clashes.

7 Conclusions

It seems as though some problems, exemplified by the 'sequence of cliques' class, are innately difficult for a whole variety of approaches. Simple statistics such as edge density are no guide to difficulty, and it is even possible that a variety of algorithms may be able to solve a certain problem easily but fail when used on a sub-problem of it. It is also clear that GAs which use the direct representation are prone to failures of co-ordination of the effort expended on solving different parts of the problem. None of the standard GAs can backtrack, and their innate parallelism can therefore cause problems. Such criticism is hardly new – for example, Gordon, Bohm and Whitley [15] observe that a GA using a direct encoding performs very poorly on large 0-1 knapsack problems, although without discussing why.

We speculate that some other GA encodings also suffer from such problems. After all, a GA can often find a good solution with substantial effort purely because mutation conducts an extensive random search, in the later stages of a run when the population has substantially converged and crossover no longer has much effect. Many GA-based papers still fail to examine whether crossover is doing anything useful in the problems they consider.

However, all this naturally suggests a possibly worthwhile direction for time-tabling research involving GAs. We suggest that a GA might be better employed in searching for a good algorithm rather than searching for a specific solution to a specific problem. Successful examples of such an approach already exist in scheduling: for example, the Philips FCM-1 robot, whose purpose is to put electronic components onto circuit boards, is configured by a GA that searches a space of parameterised greedy algorithms [24]. The speed and relative success of the many variants of sequential graph-colouring algorithms suggest that there might be versions which are very good for a given problem. Perhaps there are even variants which are very good for a whole class of problems. Even genetic programming might be useful, to search for good queue-prioritising algorithms.

8 Technical note

The reader who wishes to play about with variants of the Brelaz algorithm can find C source code at `ftp://ftp.dai.ed.ac.uk/pub/peter/`.

The file `brelaz.tar.gz` (or the equivalent `brelaz.tgz` for PC users) contains two programs:

- *brelaz*: which runs both the conventional and modified algorithms described earlier, and can backtrack if necessary. The input conforms to the syntax used by the PGA genetic algorithm package (in `pga31.tgz` or `pga-3.1.tar.gz`). By default, the output is a convenient highly-condensed representation: a timetable with N exams is represented as a string of N characters. If the i-th character is C then exam i (numbered from 0 upward) is assigned to slot number (C-'0') (slots are also numbered from 0 upward).
- *ttdecode*: this is a filter which accepts the above representation and outputs a more readable timetable, together with information about clashes, near-clashes and seats needed per slot.

The file `clique.tar.gz` or equivalent `clique.tgz` contains a program named *clique* which produces some useful information about a graph. The input format is simple: the first line should contain an integer giving the number of vertexes (assumed to be numbered from 0 upward), and all later lines define edges or cliques. For example, the line:

11 3 7 21 16

defines ten edges, one between each pair of nodes mentioned. Non-numeric material is ignored. The program can output information about edge density, node degrees, graph components, maximal cliques and can even output a list of every non-dominated clique in the graph – that is, every clique not contained in any larger clique. To do so it uses a simple brute-force strategy, although not quite so simple as to enumerate every possible subset! It can take quite a long time to run. For example, to establish that the KFU problem contained 12745 non-dominated cliques of size 3 or larger, including two of maximal size 17, took nearly two CPU-weeks on a twin-processor SPARC-20, and over 500Mb of memory.

PC users may also want the programs tar.exe and gzip.exe to unpack the above programs, eg:

```
> gzip -d brelaz.tgz
```

uncompresses the file, replacing it with brelaz.tar. Then

```
> tar xvf brelaz.tar
```

extracts all the files to a new directory named brelaz which it will create in the current directory. Run these programs without arguments to get help on their use.

All the programs have been compiled and run successfully using gcc, on PCs and on Unix workstations.

9 Acknowledgements

The second author is supported by a UK EPSRC grant GR/L22232

References

1. Michael G.Norman Ben Paechter, Andrew Cumming and Henri Luchian. Extensions to a memetic timetabling system. [4].
2. Daniel Brelaz. New methods to color the vertices of a graph. *Communications of the ACM*, 22(4):251–256, April 1979.
3. E. Burke, J. Newell, and R. Weare. A memetic algorithm for university exam timetabling. In *The Practice and Theory of Automated Timetabling* [4], pages 241–250.
4. E. Burke and P.M. Ross. *The Practice and Theory of Automated Timetabling*. LNCS 1153. Springer-Verlag, Heidelberg, October 1996.
5. M. Carter, G. Laporte, and S. Y. Lee. Examination timetabling: algorithmic strategies and applications. *Journal of the Operational Research Society*, 47(3):373–383, 1996.
6. A Colorni, M Dorigo, and V Maniezzo. Metaheuristics for high-school timetabling. *Computational Optimization and Applications*, 9(2), 1998. in press.
7. Dave Corne and Peter Ross. Some combinatorial landscapes on which a genetic algorithm outperforms other stochastic iterative methods. In T. Fogarty, editor, *Evolutionary Computing: AISB Workshop, Sheffield 1995, Selected Papers*, LNCS 993. Springer-Verlag, 1995.

8. Dave Corne and Peter Ross. Peckish initialisation strategies for evolutionary time-tabling. [4], pages 227–240.

9. Dave Corne, Peter Ross, and Hsiao-Lan Fang. Ga research note 7: Fast practical evolutionary timetabling. Technical report, University of Edinburgh Department of Artificial Intelligence, 1993.

10. Dave Corne, Peter Ross, and Hsiao-Lan Fang. Fast practical evolutionary time-tabling. In Terry C. Fogarty, editor, *Selected Papers: AISB Workshop on Evolutionary Computing, Lecture Notes in Computer Science No 865*, pages 250–263. Springer Verlag, 1994.

11. Dave Corne, Peter Ross, and Hsiao lan Fang. Evolutionary timetabling: Practice, prospects and work in progress. In P. Prosser, editor, *Proceedings of 13th UK Planning SIG*. University of Strathclyde, 1994.

12. K. Dowsland. Genetic algorithms – a tool for or? *Journal of the Operational Research Society*, 47(3):550–561, 1996.

13. A. Erguöl. Ga-based examination scheduling experience at a middle east technical university. [4], pages 212–227.

14. Larry J. Eshelman. The chc adaptive search algorithm: How to have safe search when engaging in nontraditional genetic recombination. In G. Rawlins, editor, *Foundations of Genetic Algorithms*, pages 265–283. Morgan Kaufmann, 1991.

15. V. Gordon, A. Bohm, and D. Whitley. A note on the performance of genetic algorithms on zero-one knapsack problems. Technical Report CS-93-108, Colorado State University Dept of Computer Science, 1993.

16. Tad Hogg, Bernado A. Huberman, and Colin P. Williams. Phase transitions and the search problem. *Artificial Intelligence*, 81(1-2):1–15, 1996.

17. Hugo Terashima Marin. A comparison of ga-based methods and graph-colouring methods for solving the timetabling problem. Master's thesis, Department of AI, University of Edinburgh, 1994.

18. Patrick Prosser. An empirical study of phase transitions in binary constraint satisfaction problems. *Artificial Intelligence*, 81(1-2):81–109, 1996.

19. Peter Ross and Dave Corne. Solving large multi-constrained multi-objective time-tabling problems with stochastic iterative search strategies. (journal paper, in preparation).

20. Peter Ross and Dave Corne. Comparing genetic algorithms, simulated annealing, and stochastic hillclimbing on timetabling problems. In T. Fogarty, editor, *Evolutionary Computing: AISB Workshop, Sheffield 1995, Selected Papers*, LNCS 993. Springer-Verlag, 1995.

21. Peter Ross and Dave Corne. The phase transition niche for evolutionary algorithms in timetabling. [4], pages 309–324.

22. Peter Ross, Dave Corne, and Hsiao-Lan Fang. Improving evolutionary timetabling with delta evaluation and directed mutation. In Y. Davidor nd H-P. Schwefel and R. Manner, editors, *Parallel Problem-solving from Nature - PPSN III*, LNCS, pages 566–565. Springer-Verlag, 1994.

23. Peter Ross, Dave Corne, and Hsiao lan Fang. Successful lecture timetabling with evolutionary algorithms. In A.E.Eiben, B.Manderick, and Zs.Ruttkay, editors, *ECAI-94 Workshop W17: Applied Genetic and other Evolutionary Algorithms*. ECAI-94, 1994.

24. J. D. Schaffer and L. J. Eshelman. Combinatorial optimization by genetic algorithms: the value of the genotype/phenotype distinction. In V. Rayward-Smith, I. H. Osman, C. R. Reeves, and G. D. Smith, editors, *Modern Heuristic Search Methods*, New York, 1996. John Wiley and Sons.

25. Barbara M. Smith and Martin E. Dyer. Locating the phase transitions in binary constraint satisfcation problems. *Artificial Intelligence*, 81(1-2):155–181, 1996.
26. J. Thompson and K. Dowsland. General cooling schedules for a simulated annealing based timetabling system. [4], pages 345–363.
27. P.A. Turner. Genetic algorithms and multiple distinct solutions. Master's thesis, Department of AI, University of Edinburgh, 1994.

Experiments on Networks of Employee Timetabling Problems

Amnon Meisels and Natalia Lusternik

Dept. of Mathematics and Computer Science
Ben-Gurion University of the Negev
Beer-Sheva, 84-105, Israel
Tel: +972-7-6461622; Fax: +972-7-6472909
Email: {am,natalia}@cs.bgu.ac.il

Abstract. The natural representation of employee timetabling problems (ETPs), as constraint networks (CNs), has variables representing tasks and values representing employees that are assigned to tasks. In this representation, ETPs have binary constraints of non-equality (mutual exclusion), the networks are non uniform, and variables have different domains of values. There is also a typical family of non-binary constraints that represent finite capacity limits. These features differentiate the networks of ETPs from random uniform binary CNs. Much experimental work has been done in recent years on random binary constraint networks (cf. [10, 11, 9]) and the so called phase transitions have been connected with certain value combinations of the parameters of random binary CNs.

This paper designs and experiments with a random testbed of ETPs that includes all of the above features and is solved by standard constraint processing techniques, such as forward checking (FC) and conflict directed backjumping (CBJ). Random ETPs are characterized by the usual parameters of constraint networks, like the density of constraints p_1. One result of the experiments is that random ETPs exhibit a strong change in difficulty, as measured by consistency checks, (a phase transition). The critical parameter for the observed phase transition is the average size of domains of variables. Non binary constraints of finite capacity are part of the experimental testbed. An enhanced FC-CBJ search algorithm is used to test these random networks and the experimental results are presented.

Key Words: Employee Timetabling; Constraint Networks; Experimental CSP; Non binary constraints.

1 Introduction

Employee timetabling problems (ETP) arise in the real world, out of the need to schedule employees to tasks, and form an important subfamily of timetabling problems. ETPs can be represented as a constraint network (CN) by representing tasks as variables and representing the employees as values, that have to be assigned to the variables. Domains of values for variables consist of *lists of employees* that can be assigned to the tasks that are represented by these variables.

A constraint network (CN) consists of a set $S = \{N_1 \ldots N_k\}$ of k variables, each with domain D_i of possible values, and a set of constraints among these variables, $C_{N_1} \ldots C_{N_s}$, each defined over a subset of variables $N_1 \ldots N_s$ [15]. A constraint C_L over a set of variables $L = \{N_1 \ldots N_p\}$ is defined to be a subset of the cartesian product of the domains of these variables $C_L \subseteq D_1 \times \ldots \times D_p$ [15, 6]. A *binary* CN is the case where all constraints are over pairs of variables. Values that belong to the domain D_i of a variable N_i in a constraint network, can be assigned to it, $V(N_i)$, $(V(N_i) \in D_i)$. A consistent assignment of a constraint network is a set of assignments to all variables in the network, such that all constraints between variables are satisfied [15, 5].

Binary constraints that are generic to ETP constraint networks are those that exclude employees from being assigned to more than one task at the same time. This gives rise to binary constraints of the *non-equal* (or mutual exclusion) type [3]. Employees' personal demands not to work at some tasks (or working-place constraints that forbid some types of employees to work at some tasks) rule out values from variables' domains. Therefore, domains of typical ETP constraint networks vary among variables.

A characteristic feature of ETPs is the existence of *non-binary constraints*. Typical non binary constraints are related to the fact that employees have certain limits on their working hours (number of hours per week or number of consecutive hours, for example). The structure of constraint networks that represent ETPs is discussed in detail in section 2, based on real world problems.

One major goal of the present investigation is to study the behavior of constraint networks of ETPs and in particular their difficulty. To study ETPs experimentally a testbed of random networks for ETPs is presented in section 3. Constraint networks that represent ETPs have non equality binary constraints, different domains of values for different variables, can have finite capacity constraints and have a non uniform structure. Practically all of the experimental investigations of constraint networks in recent years has focused on binary random uniform networks, with equal domains for all variables (cf.[11, 9]). The critical parameter of the phase transition that was discovered in all of these studies was a combination of constraint density and constraint tightness [11]. Since the binary part of ETPs has constraints of inequality, the tightness is very low and the interesting question is whether a phase transition occurs for ETP networks and if so, what is the critical parameter.

Tests were performed on randomly generated ETP constraint networks in order to understand the difficulty of these CNs and its dependence on their relevant parameters (section 4). The tests used search algorithms from the literature, such as forward checking and conflict-directed backjumping [10, 11]. The results of these experiments point to well defined families of ETP CNs that are much harder than others. The main result of section 4 is that the problems change from being insolvable to solvable at a certain value of average size of domains. At that value, the difficulty of ETP networks grows by large factors, relative to other values of average size of domains. To the left of the peak almost all problems are insolvable, while to the right of it most of them are solvable. The phenomenon of an abrupt change from almost all problems being solvable to almost none being solvable, is present also for different types of search techniques such as genetic algorithms (GA) (section 4.3) [12].

Strong changes in difficulty of constraint satisfaction problems have been conjectured to correlate with some critical parameter [2]. Many researchers take the view that the critical parameter of CSPs correlates with the number of solutions of the problem. Section 5 presents a simplified calculation of the expected number of solutions in our ETP CNs and demonstrates that this number decreases dramatically together with the average domain size.

2 Employee Timetabling Constraint Networks

An ETP consists of a set of tasks (shifts) for employees, each having a start time and an end time and in general these tasks may overlap in time. ETPs have a set of employees that have to be assigned to the set of tasks while satisfying all relevant constraints. Employees belong typically to a certain type and may fulfill strictly appointed tasks. The variables of the constraint network are tasks. Each variable requires a single employee (value) to be assigned to it. The values in the domains of variables are determined by two criteria

1. the legality of assigning certain employees to certain tasks.
2. some personal preferences of employees.

As a result, domains of variables of CNs that represent ETPs are in general not equal.

The most characteristic binary constraints in employee timetabling problems arise from the fact that an employee cannot be assigned to two tasks that overlap in time. Another example of a typical binary constraint is the set of requirements for a break between assigned tasks for every employee. This family of requirements excludes all tasks within the given time interval from the assignment of the same employee (value). From the above examples it is clear that binary constraints in ETPs are of the *non-equality* type (also called *mutual exclusion* in [3]). Such constraint networks are equivalent to the *list coloring* problem which is similar to graph coloring with non uniform domains. List coloring problems have been shown to be hard problems (cf. [8, 1]).

The variables that are connected via mutual exclusion constraints tend to be divided into distinct groups (those that overlap in time, for example). The connectivity of each group is very high while the number of binary constraints between groups is relatively small. Thus, employee timetabling problems have a non-uniform constraint network.

A typical *non binary* constraint of ETPs is caused by limits on employees' working hours. These include restrictions on the number of tasks an employee can work during some period (a week, for example). Constraints that limit the number of assignment of certain values to all or part of the variables are naturally called *finite capacity* constraints. These limits are non binary constraints because they connect all variables that have the corresponding employee as a value in their domains. One simple way to represent a finite capacity constraint is to use a counter [14]. Each counter represents the limit on some employee and counters are updated each time their corresponding employee is assigned to one of the variables that are connected by the counter. An assignment is compatible with finite capacity constraints if and only if the counter is not "full". In other words, as long as the value of the counter is less than its defined limit.

Many search algorithms have been devised for solving CSPs. Some of the most successful to date include forward checking (FC) and conflict-based backjumping (CBJ) (cf. [15, 10]). The standard format of the combined (FC-CBJ) search algorithm [10] can be augmented, to make it work with counters, in the following way:

- When the current instantiation of the $i - th$ variable to the value j ($i \leftarrow j$) is checked against future variables perform the following steps:
  1. For every future variable m and for every value v in the domain of m check if v is not eliminated by binary constraints between i and m. If it is, add i to the list of conflicts of m and update the domain of m.
  2. Check if the current instantiation ($i \leftarrow j$) makes the counter of j reach its limit.
  3. if it does, do the following
     (a) Eliminate the value j from the domains of all future variables.
     (b) Add all variables that were instantiated to the value j (together with i) to the lists of conflicts of the variables whose domains were updated.
  4. If the current instantiation $i \leftarrow j$ is consistent with all future variables, then:
     update the counter of the value j, i.e. increase it by 1.
  5. Move on to the next variable (i.e. $i + 1$)
- In case of *backjumping* from some variable m to a variable i perform the following updates:
  1. decrease all counters of *values* that are assigned to the variables $i \leq j \leq m$
  2. uninstantiate all variables $i \leq j \leq m$
  3. update the lists of conflicts, and domains, of all future variables of i by erasing all conflicts that involve the set of variables that have been uninstantiated.

The rest of the steps of the FC-CBJ algorithm remain the same as in [10]. In particular, back jumping from a dead end is performed to the most recent conflicting variable in the set that includes both the binary and the non binary conflicts.

Based on the above characteristic features of employee timetabling problems, the constraint networks that represent ETPs differ from uniform random CSPs in four major features

1. Binary constraints are of the non-equality type only.
2. Domains of variables are not equal and have a range of sizes.
3. ETPs have non-binary constraints of finite capacity.
4. ETP constraint networks have a non-uniform structure.

One approach to investigate the behavior of ETPs is to create a testbed for ETP constraint networks, based on some real world examples, and to test the behavior of constraint processing algorithms on ETP CNs.

The next section will address the problem of generating random constraint networks that have all of the above characteristics and define their parameters.

3   A Random Testbed for ETPs

The parameters that are needed in order to generate random networks are determined by the characteristic features of ETPs. Let there be g groups of variables in the network (think of them as the tasks that overlap in time) and let p_1 be the overall density of the (binary) constraints in the network. Each of the n variables in the network participates in N_i binary constraints *on the average*, with variables in the same group and N_o constraints, on the average, with variables in other groups, on the average *per group*. These are average numbers per variable in the network. An intuitive parameter of *uniformity* is $unif = \frac{N_o}{N_i}$. The parameter $unif$ has a range of values between 0 and 1 that is dependent on the values of p_1, g, n. In particular, for every p_1, g, n there is a *minimal value* of $unif$. For this particular value of $unif$ every group of variables is a complete network and the number of edges between groups increases with p_1 for fixed g and n. $Unif = 1$ for completely uniform networks (in which each variable is connected equally inside and outside its group).

There is one more important characteristic of ETP constraint networks, the distribution of domains of values of variables. To simplify matters we have chosen to generate the domains of variables by assigning values (employees) randomly to variables (tasks), in analogy to real life ETPs. This is equivalent to randomly selecting the subsets of employees that can be assigned to every task. The resulting domains are characterized by their sizes and the intersection of their values. To characterize the fact that different variables have different domains, we define a new parameter: the probability that a value belongs to a domain of a variable. The intuitive meaning of the parameter, which we denote *avg_size*, is the "degree of filling" of domains of variables.

Random ETP networks are characterized by $\langle n, k, g, p_1, unif, avg\_size \rangle$. The testbed of random ETP constraint networks that are created in the present investigation has a fixed number of variables n, a fixed number of possible values

k, a fixed number of groups g, and two parameters are varied in the experiments: the density of constraints p_1 and the average sizes of domains avg_size (which correlate with intersections between domains of variables). The uniformity is simply taken as the minimal for each p_1 and g has a fixed value of 5. This creates a set of random ETP networks that are characterized by the pair $\langle p_1, avg\_size \rangle$, where $0.3 \leq avg\_size \leq 1$ in our implementation.

Binary constraints of inequality rule out only a small fraction of all possible pairs of values (for k values this fraction is $\frac{1}{k}$). This fraction is denoted p_2 in the CSP literature [13] and is called "tightness" [7]. For ETPs with a large number of possible values, either employees in ETPs or colors in GCPs, the tightness is low.

Random ETPs differ from graph coloring problems by having non equal domains for their variables, as well as having non uniform networks. The first step in our experiments is to compare ETPs with equal domains with GCPs for the same parameters n,k,p_1. Figure 1 plots the median search efforts for three types of random problems with the same values of parameters ($< 30, 10, p_1, 0.1 >$).

The experiment is parameterized by the value of p_1, the density of constraints. All three types of problems behave very similarly for $p_1 \leq 0.6$ (Figure 1) and all of them are solvable. For $p_1 = 0.7$, BCSPs remain easy and solvable, but the difficulty of GCPs and ETPs jumps sharply and unsolvable problems appear. The explanation is the following: the value of tightness $p_2 = 0.1$ corresponds to easy solvable BCSPs for all values of the density p_1. For GCPs the *connectivity of the network*, which is the average number of binary constraints per variable (approximated by $p_1(n-1)$), determines the difficulty of the problems. The results of figure 1 can be taken to mean that for CNs with equal domains for all variables the uniformity of the network (which is the main difference between our GCPs and ETPs here) does not play a major role in changing the difficulty of solving the network.

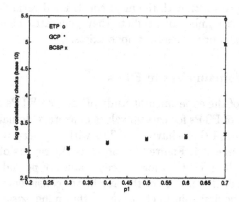

Fig. 1. Median search effort as a function of p_1 for ETPs GCPs and BCSPs

4 Experiments on ETP Networks

Experiments with random (uniform) binary CSPs have shown some pronounced changes in their difficulty [10, 13]. Random binary constraint networks are parameterized by their density of constraints (the probability p_1) and the tightness of these binary constraints which is expressed by the probability for conflicts p_2 (cf. [11]). Experiments of the last years demonstrated that for binary constraint networks with a given set of parameters $\langle n, k, p_1 \rangle$ there is a critical value of p_2 where the maximal search effort is observed. At this value a transition from solvable to unsolvable problems takes place.

Employee timetabling problems can be represented by constraint networks that have binary constraints of non-equality, different domains of values for variables, non uniform structure and some specific non binary constraints of finite capacity. Based on these observations, a set of random ETP networks was constructed. The random ETP networks are parameterized by their density of binary constraints p_1 and the average size of domains of variables avg_size. Non binary constraints of finite capacity were added to these networks and the forward checking and conflict-directed back jumping (FC-CBJ) search algorithm was enhanced to encompass these non binary constraints.

In our experiments $n = 30$. Since domains of values are different for different variables, the value k is taken to mean the total of possible values for all variables and is equal to 10. The density parameter p_1 is varied from 0.2 to 0.7 and experiments are presented graphically as functions of p_1. Every point in the $< p_1, avg\_size >$ plain of our experiments represents the average of generating and solving 100 random problems.

In order to arrive at reasonable run times for the ETP networks smart search algorithms were used. The one used in all of the experiments is a combination of forward checking and conflict-directed back jumping (FC-CBJ in [10]). The ordering of variables was static and used the *greedy largest first* strategy (orders variables according to width with the most constrained variable first) and values were ordered by the number of conflicts they generate. The search algorithm finds the first solution or proves that none exists.

4.1 Changing domain sizes in ETPs

The central result of the experimental study of random ETPs is the pronounced change in difficulty of ETPs for certain values of average domain size (avg_size). We have seen in figure 1 the behavior of ETPs with $avg\_size = 1$. Let us move on to the case of $avg\_size < 1$. Figure 2(a) presents the results of running the FC-CBJ algorithm on the ETP CNs for different values of p_1 and $0.3 \leq avg\_size \leq 1.0$. The logarithm of the median of the number of consistency checks that were performed to find the first solution or to prove that none exists is plotted against average domain size.

In view of the strong peak in solving difficulty of problems, it is interesting to calculate the fraction of solvable problems. The fraction of solvable ETP networks is plotted against the avg_size in figure 2(b) for the same set of values

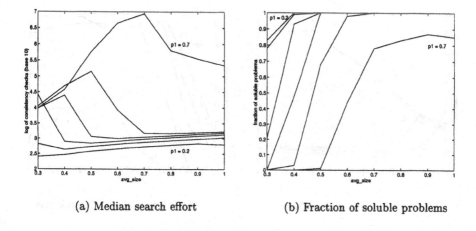

| (a) Median search effort | (b) Fraction of soluble problems |

Fig. 2. Random ETPs with binary constraints only

of p_1 as in figure 2(a). The fraction of soluble problems decreases together with *average domain size* and for $p_1 \geq 0.5$ it reaches 0 for the small values of *avg_size*. The peak of search effort occurs in the place where there is a mixture of solvable and unsolvable problems.

The results in figure 2 can be understood intuitively as follows. Let's look, for example, on the problems with $p_1 = 0.7$. When domains are small and each variable participates in a great number of constraints (p_1 is large), then the probability, that an instantiation $i \leftarrow j$ will be inconsistent, is high. Because there is a little number of values in each domain, the chance to find an assignment to the variable i is small. Therefore, most of the problems are without solution and this may be checked very fast.

4.2 ETPs with non-binary constraints

Adding finite capacity constraints to ETP constraint networks has a very clear intuitive meaning. Add limits on the number of assignments allowed for each value (employee). These limits can be implemented as counters and incorporated into the search algorithm, as explained in section 2. For our random ETPs we chose to add counters for all values. The limit on the number of assignment per employee (value) was chosen as the minimum number that does not rule out a solution. For n variables and k values the limit on all the counters is chosen as $\lceil n/k \rceil$ (same for all values). Adding this non binary constraint to the ETPs with $n = 30$ and $k = 10$ produces counters limited to $\lceil 30/10 \rceil = 3$. The results of solving these ETPs (with finite capacity constraints) are plotted in figure 3 and are similar to the results of figure 2. The shape of the graph is similar to the problems without counters. But the median search effort increases and the fraction of solvable problems, is in general lower in figure 3(b) than in figure 2(b).

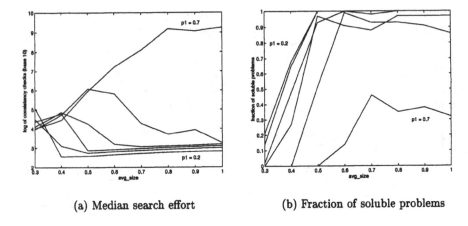

(a) Median search effort   (b) Fraction of soluble problems

Fig. 3. Random ETPs with finite capacity constraints

4.3 Solving ETPs by genetic algorithms

It is of interest to see whether the phenomenon of Figures 2- 3 depends on the type of search algorithm. Recently, there were some reports on changes in difficulty in timetabling problems when genetic algorithms (GA) were used [12]. We chose a genetic algorithm that can be described as follows:

1. Generate a population of candidate complete assignments (chromosomes). A value of each variable is called a gene.
2. Choose randomly one "chromosome" (complete assignment)
3. Perform a min-conflict mutation of it. For each randomly chosen gene (assignment) make the following steps:
   (a) check if this assignment violates some constraints.
   (b) if not, then pass to another gene.
   (c) if it is in conflict with some other genes (assignments), then
       – find another value for the corresponding variable, that participates in the minimum number of conflicts with other variables.
       – replace the current gene with the new value(gene)
4. Replace the "least fit" chromosome in the population with the new one, if the last is fitter than the first. The fitness of a chromosome i is calculated using an evaluation (fitness) function, which is inversely proportional to the number of constraints violated (see [4] for the general form of an evaluation function).
5. If there is no chromosome with the value of fitness function equal to 1 and the time is not expired, then goto 2.

We generated random ETP problems and counted the fraction of them, for which a solution was found within 3 minutes of running the genetic algorithm. The results in figure 4 show the same behavior of difficulty with average domain size as in Figure 2. The fraction of problems for which solution was found increases together with avg_size.

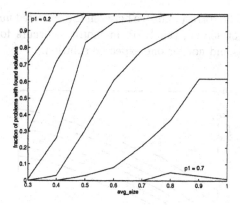

Fig. 4. Fraction of ETPs, for which solution was found (in 3 cpu minutes), *using a genetic algorithm*

5 Expected Number of Solutions of ETPs

It is interesting to calculate the expected number of solutions to see how it correlates with the presented results. To calculate the number of solutions in a list colouring problem we first express the probability of a conflict across a constraint in it.

Let C be the number of constraints. Let, Z_i be the number of zeroes (conflicts) in the constraint i. The constraint i is between two variables x and y. Let, D_x and D_y be the sizes of the domains of these variables, respectively (here each domain contains a different number of values). Then, the probability of conflict in the constraint i is defined as

$$p_{2_i} = \frac{Z_i}{D_x * D_y} \tag{1}$$

and the formula for the evaluation of the value of p_2 for a list coloring problem:

$$p_2 = \frac{\sum_{i=1}^{C} p_{2_i}}{C} \tag{2}$$

Then, the number of solutions for the list coloring problem is

$$E(N) = \prod_{x=1}^{n} D_x * \prod_{i=1}^{C} (1 - p_{2_i}) \tag{3}$$

We calculated the expected number of solutions for random ETPs (by equation 3) and the results are presented in figure 5. Clearly, the expected number of solutions decreases with decreasing *avg_size* and for large values of p_1 the number of solutions reaches 0. This approximate calculation for the number of solutions is not good enough to point to hard problems where there is exactly one expected solution as in [13]. The correspondence between the expected number of solutions and the difficulty of solving the ETP, in figure 5, occurs for small expected number of solutions and not for one expected solution.

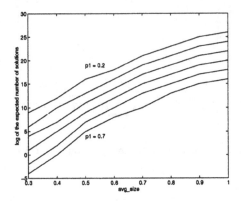

Fig. 5. The median value of the log of the expected number of solutions for random ETPs with $n = 30$ and $k = 10$

6 Conclusions

A set of experiments that included running search algorithms on random ETP networks were performed and the major result is that ETPs exhibit a strong change in difficulty together with a transition from unsolvable to solvable problems, depending on the average size of domains of variables. The peak of the search effort occurs for those values of *avg_size*, where there is a mixture of solvable and unsolvable problems. This is similar to the "phase transitions" of random CSPs that were reported in the literature in recent years [11, 13, 9]. These changes in difficulty correlate with the calculated expected number of solutions, which decreases together with the average domain size. When finite capacity constraints (which are non binary) are added to the random ETPs, the problems become more difficult (as should be expected) but behave similarly in general. All of the above phenomena does not depend on the search algorithm used.

One study of the behavior of timetabling networks was reported recently by Ross and Corne [12]. They investigated the behavior of evolutionary algorithms for solving timetabling problems and reported a pronounced peak in difficulty for certain parameters of the problems. Since their networks were defined differently than ours, the comparison of results is difficult. The reported zone of hard problems in [12] included strongly non uniform networks. The comparison of the behavior of constraint networks of different uniformities in the present study (figure 1) seems to show no great difference in difficulties for different values of *unif* (at least for $p_1 \leq 0.6$). One possible explanation for the lack of dependence on the non-uniformity of the CN is that the number of groups in the networks of Ross and Corne is small ($g = 2$), which gives rise to many more constraints inside groups, than outside. The choice of the present study is based on many real world ETPs which tend to have groups represent time periods (such as days in the week or hours in a day) and these are many more than two. In this case the number of "inside" constraints is much smaller, so, the value of *unif* is higher than in CNs with $g = 2$.

References

1. M. Arkin and E. B. Silverberg. Scheduling jobs with fixed start and end times. *Discrete Applied Mathematics*, 18:1–8, 1987.
2. P. Cheeseman, B. Kanefsky, and W. M. Taylor. Where the really hard problems are. In *Proceedings of the Twelfth International Joint Conference on Artificial Intelligence*, pages 331–337, Sydney, Australia, 1991.
3. B. Y. Choueiry and B. Faltings. Temporal abstractions and a partitioning heuristic for interactive resource allocation. In *Notes of Workshop on Knowledge-based Production Planning, Scheduling and Control, IJCAI-93*, pages 59–72, Chambery, France, 1993.
4. D. Corne, P. Ross, and Hsiao-Lan Fang. Fast practical evolutionary timetabling. *Lecture Notes in Computer Science*, 865:250–263, 1994.
5. R. Dechter. Constraint networks. In S. C. Shapiro, editor, *Encyclopedia of Artificial Intelligence, 2nd Edition*, pages 276–285. John Wiley & Sons, 1992.
6. R. Dechter and J. Pearl. Network-based heuristics for constraint satisfaction problems. *Artificial Intelligence*, 34:1–38, 1988.
7. R. Dechter and P. vanBeek. Constraint tightness and looseness versus local and global consistency. *J. of ACM*, 44:549–566, 1997.
8. M. C. Golumbic. Algorithmic aspects of perfect graphs. *Annals of Discrete Mathematics*, 21:301–323, 1984.
9. S.A. Grant and B.M. Smith. The phase transition behaviour of maintaining arc consistency. In *Proceedings of the 12th European Conference on Artificial Intelligence*, pages 175–179, Budapest, Hungary, 1996.
10. P. Prosser. Hybrid algorithms for the constraint satisfaction problem. *Computational Intelligence*, 9:268–299, 1993.
11. P. Prosser. Binary constraint satisfaction problems: some are harder than others. In *Proceedings of the 11th European Conference on Artificial Intelligence*, pages 95–99, Amsterdam, 1994.
12. P. Ross, D. Corne, and H. Terashima. The phase transition niche for evolutionary algorithms in timetabling. In *Proceedings of the 1st Conference on Practice and Applications of Automated Timetabling*, pages 269 – 282, Edinburgh, UK, August, 1995.
13. B. M. Smith. Phase transition and the mushy region in csp. In *Proceedings of the 11th European Conference on Artificial Intelligence*, pages 100–104, Amsterdam, The Netherlands, 1994.
14. G. Solotorevsky, E. Shimony, and A. Meisels. Csps with counters: a likelihood-based heuristic. In *Proc. Workshop on Non Standard Constraint Processing, ECAI96*, pages 107–118, Budapest, August, 1996.
15. E. Tsang. *Foundations of Constraint Satisfaction*. Academic Press, 1993.

Evolutionary Optimisation of Methodist Preaching Timetables

David Corne,   Revd Dr. John Ogden

Department of Computer Science, University of Reading,
Reading, RG6 6AY, UK. Email: D.W.Corne@reading.ac.uk

Abstract. Methodist churches are arranged into local *circuits* with, for example, some 30 churches existing in a local area covering about 1,000 square miles. A central element of Methodist religious life at a particular church is the frequent delivery of sermons and similar activities by preachers who are *not* based at that particular church. Periodically, it is the job of a senior member of the Methodist movement in a local area and his or her staff to draft a *preaching timetable*. This may involve, for example, about 40 or 50 ministers (of different seniorities and types), 30 or 40 churches, and the need to fill each of 3 or 4 preaching slots every Sunday at each church in the region. The problem of finding a suitable preaching timetable involves a variety of idiosyncratic constraints which make it an interesting variation on the general timetabling problem. Here we look at various simple approaches to the problem, which draw on the two main styles of approach to stochastic iterative timetabling (direct and indirect representations), and compare three well-known search techniques: evolutionary algorithms, simulated annealing, and hillclimbing. In the context of finding a successful approach to a problem with development time at a premium (hence: using simple implementations) we find strong support for the superiority of an 'indirect' timetable representation over a direct one. We also find that simulated annealing is generally the more robust method over a range of problems, with all other methods except hillclimbing performing strongly in particular cases.

1   Introduction

Methodist churches are arranged into local *circuits* with a head office for each such circuit. For example, some 30 churches existing in a local area covering about 1,000 square miles might constitute a 'circuit', and administration issues involving the circuit as a whole are dealt with in the 'circuit office' by an appointed senior minister and a small number of administrative staff.

The key issue dealt with at circuit level is the movement of methodist preachers 'around the circuit' to deliver sermons at different local churches. This activity is a central element of Methodist religious life, and leads to a quarterly 'preacher planning' problem which must be solved at the circuit office, producing a timetable of sermons at each church to be published in advance of the next quarter. This problem involves a diverse and particular collection of constraints, which render it both interestingly similar and interestingly similar to the exam and course timetabling problems. The major constraint is that each available church service must be 'filled' by a sermon, but many complicating soft constraints exist such as the need for reserve preachers, and individual ministers' preferences with regard to travel, timing, and other issues.

The current approach to this problem is manual, and involves several days of struggling with pencil and paper, and making telephone calls to negotiate with ministers about their particularly awkward preferences. There is, not surprisingly, a need for automated solutions to this problem. Better use of senior ministers' time is clearly part of the motivation for this, but perhaps the main need for good, automated, optimised preaching timetables is to enhance the attainment certain objectives in the Methodist religious system. Quite apart from making sure that elderly ministers do not need to drive too far from home in the evening, for example, a key issue is for ministers to 'get around the circuit' over a three or four year period. Ideally, every minister or lay preacher will have visited most of the churches in a circuit over such a period, with a decently long amount of time before revisiting the same church twice. Especially in conjunction with the various hard constraints, these issues are what make the problem particularly difficult.

This paper describes our approach to this problem, which essentially consisted of drawing on the two most frequent approaches to stochastic iterative timetabling (direct [3, 4] and indirect/greedy [1, 8] representation), and massaging them to cope with the problem at hand. We are thus able to report on the relative success of these two approaches when applied to a new timetabling-like problem. The search techniques we tested were a straightforward evolutionary algorithm, simulated annealing, and hillclimbing.

The paper is set out as follows. Section 1 provides further detail and definition of the preacher planning problem. In section 2 we then describe our approach to solving the problem, in terms of representation issues and the penalty function. Section 3 then describes the search methods used, and section 4 describes the test problems (which are available via the EvoStim WWW site.)[1] Experimental setup is briefly described at the start of Section 5, which then continues with an exposition of our results. Conclusions appear in Section 6.

The Preacher Timetabling Problem

The problem we are faced with begins with a partially filled in preacher plan for a given three-month period. For each sermon 'slot' in each church in the circuit, and for each Sunday in the period (there will be 12, 13, or 14 Sundays in the quarter), this plan either sets out which minister is to deliver the sermon, or indicates that no minister is yet planned for this sermon. In most cases, the existence of a minister 'filled in' for a sermon, simply means that the minister

---

[1] EvoStim is the European Network of Excellence in Evolutionary Cmputation's Working Group on Scheduling and Timetabling; EvoNet is at http://www.dcs.napier.ac.uk/evonet .

responsible for that church has already decided to give that particular sermon. On the other hand, the many empty slots are left empty precisely for the purpose of visiting ministers or lay preachers from elsewhere in the circuit. Figure 1 broadly illustrates the starting point for the problem. The grid in this figure is made up of cells each divided into three, representing the (typically) three services per Sunday at a church. For example, Figure 1 indicates that "S.Ermon" is the resident minister at church no. 3 (this is clear, since S.Ermon seems to deliver most of the sermons at this church). Consequently, most of the services are already 'filled' with S.Ermon's name. Services left blank are available for visiting preachers from elsewhere in the circuit.

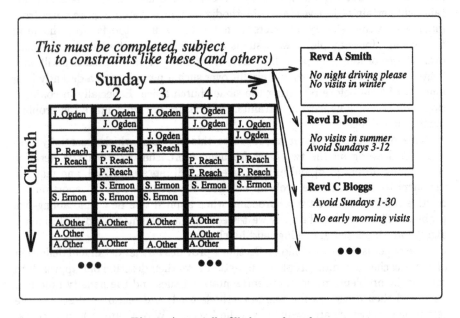

Fig. 1. A partially filled preacher plan

The problem, simply stated, is to fill all of the empty slots with a suitable minister. The constraints are as follows:

Availabilities: Each minister has expressed his or her availability for each Sunday in the quarter (usually numbered 1 to 13). Usually, a minister is available for 5 or 6 Sundays, and would be expected to preach around the circuit around 3 or 4 times.

Preferences: There is a long list of different types of preferences that preachers can express for the Sundays they have available. The main ones are:

    – Some preachers may request not to visit a church too far away in the evening, especially in winter (to avoid driving long periods in the dark).

– Some preachers want to avoid traveling to relatively distant churches on successive Sundays.

– A 'double' means delivering a sermon in the morning or afternoon, and then delivering another sermon elsewhere in the afternoon or evening of the same Sunday. Some preachers express a preference for doubles, while some express the need to avoid them.

Special Services   Certain services require an *ordained* minister, rather than an 'ordinary' minister, or a lay preacher.

Family Services   Some services are specified on the plan to be 'all age worship' or 'family services'. This is a particularly challenging situation which is best handled by a suitably experienced minister.

Training Services   When a *novice* minister is scheduled, an *ordained* minister must be present for training/assessment purposes.

Spread: Individual ministers should 'cover the circuit' over several years, so there is an objective to maximise the amount of time between visiting the same church twice. In particular, attainment of this objective needs reference to the finished preacher plans of previous quarters.

Reserves: For emergencies, there need to be preachers available in reserve who would be able to take over any particular sermon at short notice.

It currently takes several days of staff time at the circuit office to generate a preaching timetable for the ensuing quarter. In common with most manual attempts at timetabling problems, the result tends to just about manage to cover the hard constraints (all slots filled), but leaves rather a lot to be desired with respect to the softer constraints.

For example, certain ministers are cajoled into delivering more sermons away from home than they had wanted to, other ministers may need to do some unwanted night-time driving, or unwanted 'doubles'.

The second author implemented the first ever software system for generating preacher timetables for this purpose, nearly 20 years ago. This system used simple heuristics and did not much improve on the kinds of timetables manually achievable, although it greatly speeded up the process. It fell out of use several years ago as a result of changes to the preacher planning process, and never getting around to to update the system appropriately.

2   Representation and Evaluation

This section describes the simple preacher timetable solution representation method and the cost function, developed to support experimentation with evolutionary and related search techniques.

Direct Representation

A solution to the preacher planning problem is simply a choice of preacher for each unfilled sermon. This lends itself to a simple representation as a vector

P, whose length is the number of unfilled sermons, and where each element is an index into the list of the preachers available for that particular sermon. For example, "5,3,5,12,8,..." means "use preacher 5 for sermon 1, preacher 3 for sermon 2, preacher 5 again for sermon 3, preacher 12 for sermon 4," and so on.

This is a 'direct' representation, in the sense that the vector P directly translates into a candidate solution. One issue which arises about this, however, is what we mean by 'available' in the above. It turns out that if we restricted gene values for a particular locus (sermon) to range through only the preachers who have indicated availability for that sermon, it is often very difficult to find a solution which does well in terms of the other constraints. Instead, the gene values for every locus can range through 'all' preachers, which turns out to yield better solutions in terms of the other constraints, but which mean violating one or two availabilities (which can then be explained and negotiated via a telephone call)[2]. In this work, we allow a gene to range through all preachers, and hence need to apply a penalty for violation of availability.

The exception to this is the issue of Special Services (eg: Communion), and Family Services, which respectively require an ordained minister, and an experienced minister (all ordained ministers are experienced, and a few others). This is a hard constraint, so the genes which represent such sermons are indeed restricted to indexing only suitable preachers.

In conclusion, note that the need for two preachers at a training sermon seems to require modification to the simple method described. In practice, this issue is dealt with in the cost function, described below, without need for a change in the representation. It is difficult to deal with in the representation itself because we do not know in advance which sermons will turn out to be 'training' sermons. A training sermon (and hence the need for another, experienced preacher to attend) occurs whenever a novice preacher happens to get assigned to an unfilled slot.

Cost Function

The representation guarantees that all sermon slots are filled by a suitable preacher. The cost function therefore only considers the quality of the timetable in respect to the other constraints. Our approach is straightforward, and amounts to adding up a weighted sum of constraint violations. There are various complications however, which are individually described in the exposition below.

Availabilities When a preacher is assigned to a slot for which they indicate unavailability, a penalty P_u is used. However there is an extra factor here concerning the number of sermons for which a given preacher is available. For example, it is particularly unfortunate if a minister is only available for one particular Sunday slot, but we either don't use him or her at all, or assign him or her to a different slot. In other words, it seems to make sense to pay more attention to

---

[2] This and similar issues, in the context of the current manual approach to solving the problem, are the subject of continual debate at the circuit office

'barely available' preachers than to 'readily available' ones, owing to their implied relative flexibilities. To deal with this we apply a penalty $U/(1 + A)$ for each preacher, where A is the number of assignments of this preacher to sermons for which they are unavailable, and A the number pf assignments to sermons for which they are available. This is summed for each preacher, and multipled by the aforementioned penalty factor P_u.

Avoiding Long Drives: In practice, the circuit ministers and other staff who develop the quarterly plan use much special and personal knowledge of the individual preachers, manifest in proceedings such as "Oh, we'd better not send old Arthur all the way to Henley that Sunday night; he would probably prefer to stay at home in the evening that late in the quarter." We 'capture' such knowledge by simply maintaining a matrix of travel distances between each church and each preacher's home, and by storing a binary flag for each preacher which indicates whether or not that preacher would like to (or be assumed by the circuit minister to) avoid driving long distances for evening services. The upshot of this is that we can find, for each thus-flagged preacher, the distance traveled for any evening services they are assigned to. Adding up all such distances, this sum is added to the cost with a weighting factor P_d. We are not yet dealing with the seasonal issue, whereby the avoidance of long drives only really counts late in the October to December quarter, and early in the January to March quarter.

Doubles For each preacher, we store an indication of whether they like, dislike, or are indifferent to doubles. Penalties $P_l d$ and $P_d d$ are then applied, respectively, for each instance of a preacher who likes doubles but is not given one, and dislikes doubles but is given one.

Spread: We maintain on file, for each preacher, the number of weeks since they last visited each of the churches on the circuit. This of course incorporates data obtained from previous plans. A '-1' weeks indicator simply means that the preacher has never visited a particular church, while a '0' indicates irrelevance – ie: the preacher happens to be the minister 'resident' at the church in question.

In the course of evaluating a particular candidate plan, this information is used, in conjunction with the current plan, to calculate, for each church a preacher visits on the plan, the number of weeks between this visit and the last visit. In each case we either return a given number of weeks between visits, or an indicator that this church is visited by this preacher for the first time in the current plan. If a preacher does not visit this church on the plan, but was last visited by this preacher a certain number of weeks ago (the most common instance), this translates into a number of weeks between visits which assumes the preacher will visit that church roughly in the middle of the *next* quarter.

With c churches and an individual typical number v_p visits for each preacher p (which is stored), it is easy to estimate the ideal number of weeks between a preacher's visits at a particular church; this is $x_p = 13v_p/c$. We want to minimise the deviation from this ideal number for each church, and so add a penalty term $(w_p - x_p)^2$ for each preacher/church combination in which we have a valid 'number of weeks between visits'.

Reserves The cost function builds an array which details, for each sermon, all of the preachers who have not been assigned to that sermon, but who would be available to do it (ie: who have not expressed unavailability for that slot). We need to ensure that as many such preachers are available for each sermon, to maximise the chances that a reserve can be found if needed. To handle this in a 'penalty function' context we add certain terms which must be minimised. These are: the number of sermons which have no suitable reserves, the number of sermons which have only 1 suitable reserve, the number of sermons which have only two suitable reserves, and so on, up to 5 suitable reserves. The associated numbers are respectively weighted by the six penalty terms P_r01, ...,P_r5. Note that in the case of a Special Service, 'suitable reserve preacher' means a 'reserve ordained minister'. The penalty weightings obviously descend from P_r01 to P_r5, being particularly high for P_r0.

Training Services When a *novice* minister is scheduled, an experienced minister must be present for training/assessment purposes. To handle this, the cost function builds a list of experienced ministers who are still available (ie: not unavailable, or assigned elsewhere at the same time in the candidate solution) at the same time as such a service. The idea is to maximise the number of suitable preachers available to sit in. This is dealt with in a precisely analogous way to how we deal with Reserves (above), this time with five penalty factors P_t0,...,P_t4, relating to the number of such services with 0, 1, 2, 3, and 4 such preachers respectively. This enables us to maximise the choices available for 'training' preachers. This system, however, does not actually assign a particular minister to attend; the issues and factors relating to a good choice of such a minister are complex, and also take into account the particular novice preacher involved. The objective is therefore to maximise the choices, and leave it to the circuit head minister to make the final assignment in these cases.

Indirect Representation

The 'indirect' representation approach involves using a 'plan building' algorithm which builds the timetable step by step (finding a preacher for each sermon in turn) attempting to avoid conflicts and constraint violations as it goes. The quality of the end result depends on choice points within the algorithm; in the context of an iterative search method, an evolutionary or other algorithm searches through the space of choice points. A typical strategy, which we employ here, is to construct a plan builder whose choice points come down simply to deciding which sermon to fill next. The job of search is then to search through the space of orderings of sermons.

The basic plan builder we employ works as follows:

Input: A list of unfilled sermons in a particular order.

Iterate: Take the next unfilled sermon. Using the current partially built timetable, and stored constraint information, assign a score to each preacher. Assign the preacher with the best score to this sermon. In the (actually quite rare) case of a tie, simply take the first.

The key point is of course the scoring mechanism. To assign a score to a preacher for a given sermon we essentially apply the cost function detailed above, restricted (where this makes sense) only to the sermons assigned so far. Of course, scores are only 'suitable' preachers are considered in the cases of Special services and Family services.

The input to this schedule builder is a permutation of the list of unfilled services. The chromosome in this case is therefore such a permutation, and of constrained to stay that way via simple specialised operators.

3   Search Methods

Using both representations, we tested and compared four search methods on examples of the preacher planning problem. Below, we first describe certain operators common to the methods. These are not new, but are shown here for completeness. We then provide the essential details of the algorithms: hillclimbing, simulated annealing, and an evolutionary algorithm.

3.1   Operators

Single Gene Mutation This operator is used only in direct-representation cases. Given a chromosome, a locus chosen uniformly at random, and the gene at that locus is given a new value, chosen uniformly at random but not allowing replacement with the same value.

Swap Mutation This operator is used only in the indirect representation cases. Given a chromosome, choose two gene loci uniformly at random, ensuring they are distinct. The alleles at these loci are then swapped. In this application, this guarantees producing a different genotype, since the indirect representation chromosome is permutation with no repetitions. (However, there is no such guarantee for producing a different phenotype).

Uniform Crossover This is used only in the direct representation case. Standard uniform crossover as described by Syswerda [10], where a single child is produced from two parents by, for each locus in turn, taking the value at that locus from a random one of the parents.

Order-Based Crossover Order-based crossover [5] is used only in the indirect representation case. A small number of permutation loci (between 3 and 6, in this case) are chosen at random in one parent. The gene values at these loci in the first parent are noted. The child is first constructed as a copy of this first parent. The same gene values are then located in the second parent, and the order in which these genes appear in the second parent is imposed on them in the second parent (but remaining in the same set of loci as the first parent). This of course ensures that the child remains a valid permutation.

3.2 Hillclimbing

In Hillclimbing (HC), a random chromosome is first generated, becoming the current chromosome g, and is evaluated and found to have cost c. The following is then iterated a given number of times: mutate the current chromosome producing a new chromosome $g\prime$ with cost $c\prime$; if $c\prime \leq c$, g now becomes $g\prime$; if not, $g\prime$ is discarded. Hence, a chain of mutations is performed to the current chromosome, replacing the current chromosome by the latest mutant whenever the latter is fitter or equal to the former. Experiments (unless explicitly stated otherwise) always stop after 30,000 chromosome evaluations.

Simulated Annealing

Simulated Annealing [7] (SA) differs from HC in that instead of replacing g with $g\prime$ only if $g\prime$ is no worse than g, we replace g with a worse $g\prime$ with a probability $e^{(c-c\prime)/T}$. T varies during the run, starting at a relatively high value, so that most worse mutants are accepted, and finishing at a much lower value, such that worse mutants are very rarely accepted if at all. The initial and final values of T were chosen such that, to begin with, the chance of accepting a minimally worse mutant was 0.99, and to finish with this chance was 0.0001. The temperature was changed after every 500 evaluations using a constant multiplier calculated appropriately.

Evolutionary Algorithm

The evolutionary algorithm [6] (EA) employed uses local mating selection [2] with a walk length of 2, steady state reproduction, and a population size of 100. Uniform crossover and single gene mutation were employed in the direct representation case, with order-based crossover and swap mutation employed in the indirect mutation case. The GA is elitist; when a chromosome is chosen to be destroyed to make way for a new entrant, this will never be a unique current best chromosome. The chromosome chosen for destruction is replaced if the new entrant is fitter or equal to it, or with probability 0.1 if not.

4 Test Problems

Abstracting from the general nature of the problem as described above, we have built a generator for preacher planning problems of this type which will be made available along with code for evaluation. This will hopefully be useful for comparative research, and is also designed specifically to assess the suitability of our methods for different circuits. Worldwide, the essential elements of the Methodist preacher planning problem remain the same, but various differences such as the ration of required visits to available preachers, number of churches in the circuit, distances between churches, and so on, may well make major differences to the fitness landscapes (and hence the relative superiority of different search methods) in each case.

The test problems we used in the current study were five different problems designed to emulate the constraints and features of the Reading area circuit. Each involves 15 churches, 30 preachers, and a realistic collection of constraints based on 'typical' parametrisations which are given as input to the problem generator. These parameters include such things as: the proportion of ministers who are ordained, the proportion of unfilled services, the proportion of ministers who like doubles, and so on.

5 Experiments

In all, six different algorithms were tested on each problem, resulting in 30 experiments (2 representations × 3 algorithms × 5 problems). Each such experiment involved 100 trials with different random seeds. The results collected were the best cost found in each trial, and the key result for an experiment (set of 100 trials) is the best, worst, and mean such best results.

We summarise the results here by indicating relative performance graphically with respect to one of the test problems, which displayed quite typical behaviour. Full results (best, worst, and mean values for each of the 30 experiments on each of the 5 problems) are available at the indicated WWW site along with the problem generator and evaluation function.

First, Figure 2 indicates comparative performance of the three algorithms when using the direct representation.

Clearly, and as indicated by the summary of the statistical test indicated in the figure, simulated annealing turned in the superior performance, significantly improving on hillclimbing, and with the GA performing far worse than either other method.

Figure 3 summarises results on the same problem when using the indirect representation approach.

Again, the GA turns in the worst performance, but first and second place are now solidly reversed. Simple hillclimbing is now the better method. It is worth comparing the direct and indirect approaches directly; this is done for convenience in Figure 4.

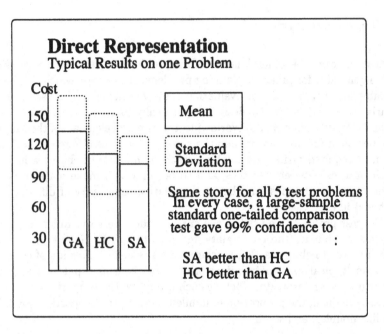

Fig. 2. Performance Histogram on one problem using the Direct Representation

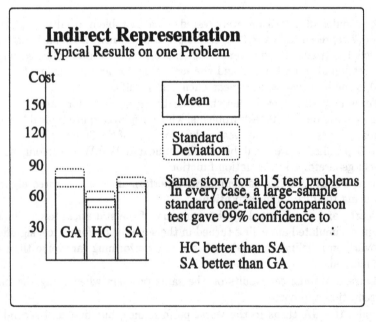

Fig. 3. Performance Histogram on one problem using the Indirect Representation

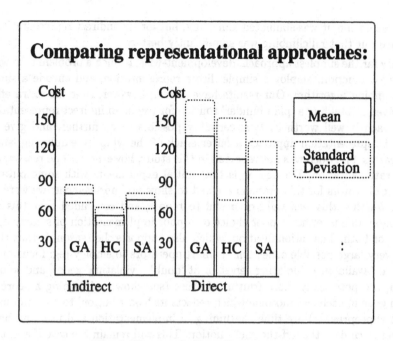

Fig. 4. Direct vs Indirect Representation

The vertical axes correspond in the two parts of Figure 4, so we can clearly see that even the best performance when using the direct representation (simulated annealing) is defeated by the worst performance when using the indirect representation (the genetic algorithm). Generally, this is quite strong evidence for the superiority of the indirect representation approach, although with certain important qualifications which shall be mentioned in the section 6.

6 Conclusions

This paper has looked at an interesting problem in the timetabling area, comparing some common approaches which tend to be taken to optimisation in this field. The general spirit of this work was that of a rapid application development; in this sense, rather than work hard at optimising any particular approach or technique, we treated the techniques we tried as established 'off-the-shelf' remedies to the need for automated good-quality preacher timetable optimisation.

Thus taking the six algorithms off the shelf (3 established methods × 2 established representation techniques), we found the combination of hillclimbing and indirect representation to be the most fruitful. This was strongly confirmed in statistical terms. Other general findings were that the indirect representation was much better than the direct representation overall, and that the identity of the best performing method switched between representations. For the direct representation, it was simulated annealing, but for the indirect representation it turned out that hillclimbing was significantly better.

By far the simplest approach, development-wise, to such a problem is to write the cost function, employ a simple direct representation, and encode a simple hillclimbing algorithm. Our results have shown, however, that the extra effort involved in building a 'plan builder' routine for use in an indirect representation approach is well worth it. It is certainly possible to go further, and give the direct representation approach a fairer crack of the whip by employing 'smart operators [9]' . Such was not reported in this study, however, for one reason, and one rationalisation. The reason is that initial experiments with some putative smart operators for this problem seemed to show no improvement; we were not sure whether this was valid or useful to report, and concluded that this was probably due to either a poor choice or a poor implementation of smarty operator, or both. The rationalisation, which also supports the reason, is that there is a very large possible collection of smart operators to employ (eg: focus mutation at 'availability' violating genes, or at 'doubles violating' genes, and so on.). Also, the potentially more fruitful of these (somehow maintaining a score for each gene in each chromosome which reflects its 'contribution' to cost and using that appropriately) are time consuming in implementation and use, and hence rather invalidate the 'off-the-shelf' notion. This will remain the case if and until an established methodology arrives for good practice in choice and implementation of smart operators for direct representations.Hence, what we have not really shown, is whether or not further extra effort at employing 'smart operators' in the direct representation may re-establish competitive performance for this technique. But we have shown that, in the case of the preacher timetabling problem, the indirect representation is clearly better than a simple 'direct representation' approach.

An interesting finding is the reversal in fortunes for simulated annealing and hillclimbing as we switch between representations. This clearly seems to indicate aspects of cost landscape structure in this problem, which may also fuel thoughts into the reason for a GA generally performing so poorly.

As ever, in such cases, the 'poor' performance of the GA is probably best understood by the good performance of HC and SA. That is, the landscapes involved seem to be amenable to the assumptions in HC and SA, namely that gradient descent, with some allowance for 'bumps (in SA's case) is a good heuristic. In other words, the landscape structure, especially in the indirect representation case, was either unimodal (which seems unlikely) or multimodal with steep-sided optima. In the latter sort of case, HC or SA will rapidly climb towards one of the many optima, whereas the GA will waste much time exploring Hamming-distant leaps from the current point, which would be better spent (as do HC and SA) in nearby exploitation of current points.

The HC/SA switch in performance between direct and indirect representation seems to indicate a smoother landscape in the indirect representation case. This seems to make sense: it is intuitively clear that, for example, very good points in the direct representation are very close to poor points (bumps) which arise from changing the value of one gene (for example, to that of a preacher ostensibly unavailable for that service). However, 'close' mutations in the indirect representation rarely have such a direct effect. The operation of the plan builder is generally such as to do the best it can given the current input order of the sermons. A slight change in this ordering does not affect the general bias towards avoiding conflicts and poor choices.

Finally, there is much interesting work to do in this problem area. Because Methodist preaching circuits exist worldwide, there is a great variety of problems of this general kind which differ in such things as the numbers of churches and preachers involved, and the general pattern of constraints. It may be the case, for example, that the performance differentials we observe here are altered for different instances of such problems. This is under investigation.

References

1. E. K. Burke, D. G. Elliman, and R. F. Weare, 'The automated timetabling of university exams using a hybrid genetic algorithm', in *AISB Workshop on Evolutionary Computing (University of Leeds, UK, 3-7 April 1995), Society for the Study of Artificial Intelligence and Simulation of Behaviour (SSAISB)*, (1995).
2. Robert J. Collins and David R. Jefferson, 'Selection in massively parallel genetic algorithms', in *Proceedings of the Fourth International Conference on Genetic Algorithms*, eds., R.K. Belew and L.B. Booker, pp. 249–256. San Mateo: Morgan Kaufmann, (1991).
3. Dave Corne, Hsiao-Lan Fang, and Chris Mellish, 'Solving the module exam scheduling problem with genetic algorithms', in *Proceedings of the Sixth International Conference in Industrial and Engineering Applications of Artificial Intelligence and Expert Systems*, eds., Paul W.H. Chung, Gillian Lovegrove, and Moonis Ali, 370–373, Gordon and Breach Science Publishers, (1993).
4. Dave Corne, Peter Ross, and Hsiao-Lan Fang, 'Fast practical evolutionary timetabling', in *Proceedings of the AISB Workshop on Evolutionary Computation*, ed., Terence C. Fogarty, Springer-Verlag, (1994).
5. B.R. Fox and M.B. McMahon, 'Genetic operators for sequencing problems', in *Foundations of Genetic Algorithms*, ed., J. E. Gregory, 284–300, San Mateo: Morgan Kaufmann, (1991).
6. David E. Goldberg, *Genetic Algorithms in Search, Optimization & Machine Learning*, Reading: Addison Wesley, 1989.
7. S. Kirkpatrick, C.D. Gelatt, Jr., and M.P. Vecchi, 'Optimization by simulated annealing', *Science*, **220**, 671–680, (1983).
8. B. Paechter, A. Cumming, H. Luchian, and M. Petriuc, 'Two solutions to the general timetable problem using evolutionary methods', in *Proceedings of the First IEEE Conference on Evolutionary Computation*, pp. 300–305, (1994).
9. Peter Ross, Dave Corne, and Hsiao-Lan Fang, 'Improving evolutionary timetabling with delta evaluation and directed mutation', in *Parallel Problem Solving from Nature III*, ed., Y. Davidor, Springer-Verlag, (1994).
10. G. Syswerda, 'Uniform crossover in genetic algorithms', in *Proceedings of the Third International Conference on Genetic Algorithms and their Applications*, ed., J. D. Schaffer, 2–9, San Mateo: Morgan Kaufmann, (1989).

Improving a Lecture Timetabling System for University-Wide Use

Ben Paechter, R. C. Rankin, and Andrew Cumming

Department of Computing
Napier University
219 Colinton Road
Edinburgh
Scotland
EH141DJ
benp@dcs.napier.ac.uk

Abstract. During the academic year 1996/97 the authors were commissioned by their institution to produce an automated timetabling system for use by all departments within the Faculty of Science. The system had to cater for the varying requirements of all the departments, be easy to use, robust, expandable, and timetable 100% of events fully automatically within a reasonable time. The timetables produced had not only to be workable, but also had to be 'good' with respect to management defined criteria. The work was intended as a pilot study for later extension to the whole institution. This paper describes the enhancements to the user interface and timetabling engine that were found to be necessary to meet the more extensive needs of a faculty.

1. User Interface

Specifying timetabling constraints can be difficult for anyone, and often this task is to be completed by administration staff with little knowledge of mathematics or computing. It is important therefore, that the user interface is easy to use, and that the concepts employed related to the real world rather than to abstract mathematical structures and operations.

In designing the interface some concepts were defined to help users to specify the constraints on the timetable.

1.1 Features

A *feature* is a property *satisfied* by some resources or events in the system and *required* by others. For example certain rooms may be "accessible" by those with mobility problems. The room is made to *satisfy* a feature called "access" that is *required* by a student-group. The constraint being represented here is that any event that the student group is required to attend can only be allocated to a room possessing the "access" feature. Another constraint might be that a certain room is reserved for senior students. This could be represented by making the room *require* the feature "senior" and the senior student groups *satisfy* it.

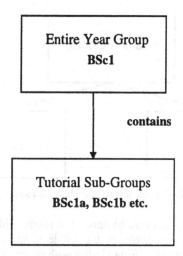

Fig. 1. The *contains* relationship for groups.

There is no restriction on what can be represented as a feature; the most common use in a computing department is to manage rooms according to function and equipment and software availability. Features need not be singular: a room may satisfy or require several features. For example, a pair of computer laboratories for data-communications classes may have the same equipment and software but one may be more accessible than the other. Both would have the feature "comms", but only the latter would also have the feature "access". Events requiring "comms" could be placed in either laboratory, but "comms" events that also required "access" could only be placed in the latter.

1.2 Containers

A *container* is a resource that can hold other similar resources, including other containers. The *contains* relationship can be seen as a *parent-child* relationship giving us a directed graph. If a particular resource is busy then all its descendants are also busy along with all of the descendant's ancestors.

This has been applied to student-groups and rooms. A common example is show in Figure 1, where a student group sub-divides for tutorial classes or practical classes.

The container concept allows for flexible use of rooms, especially when different features are associated with each level in the container hierarchy. For example, a general purpose classroom may have the feature "GP-room" and a nearby computer laboratory may have the feature "PC-lab". Although there is no physical relationship between these rooms, a container may be created that encompasses both and has the abstract feature "GP+PC". Any events of a subject or module that need flexible access to both types of accommodation will require the feature "GP+PC". These events will be timetabled into the logical container room making both physical rooms busy at that time. The composite room becomes unavailable when either of its components is in use.

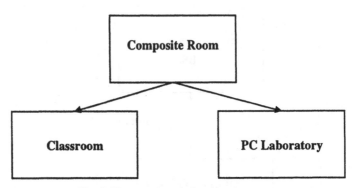

Fig. 2. The *contains* relationships for rooms

The container relationship can also be applied to rooms with moveable partitions which allows the various components, or the whole room, to be used at will but prevents booking of one element causing clashes with others. There is no limit to the depth of the container tree and its branches may be unbalanced.

A novel use of the container function is to deal with rooms of variable capacity. For example a PC laboratory may have 16 computers. For one class it may be desirable to have one student per machine, while for another, students may benefit from working in pairs. One room is created with size 16 and the feature "PC-lab" and another with size 32 and the feature "crowded-PC-lab". One room is made to contain the other. Events that allow one student to each machine will require a "PC-lab" and those that allow two students per machine will require a "crowded-PC-lab". As before if the parent is busy the child is unavailable and vice-versa, so clashes do not occur.

A more specialised use of the container facility occurs where an event requires use of different types of accommodation at different times in the semester. For example, if a room is only required by an event on odd weeks of a semester that event can be made to require "odd-ness". A physical room is made to contain two identical rooms, one with the feature "odd-ness" and the other with the feature "even-ness". Because siblings can be assigned independently, the allocation of the "odd" room to an event does not preclude the use of its sibling by an event requiring the "even" room. This technique is unnecessary where the system deals with separate timetables for each week, but is of importance where the greatly increased complexity and processing load of supporting weekly timetables needs to be avoided.

2. Evolutionary Engine

The evolutionary engine employs a memetic algorithm using an indirect representation, heuristic seeding, directed mutation and targeted mutation. This algorithm is described in detail in [1]. Changes to the algorithm, and further results are given below.

2.1 Room Assignment

In previous versions of the algorithm, genetic representations encoded information about the timeslot that each event should take place in, and rooms were allocated using a greedy algorithm which allocated the best fitting room available to each event. However, new evaluation criteria concerning the movement of people around and between buildings has made it necessary to extend this genetic representation to include information about the room in which an event should take place.

The specified room for an event is chosen from an ordered list of possible rooms. A room is deemed "possible" if the combination of event, student groups, lecturers and that room, leaves all the required features of any of them satisfied by at least one of them, and the room capacity is greater than the combined sum of the sizes of the student groups attending. The list is ordered by a heuristic that takes into account features that the room satisfies that are not required, and unused capacity of the room. This helps to ensure that events do not claim rooms that are bigger or satisfy more features than are necessary.

When evaluating a chromosome, if the room specified for an event is not available (because it has been used by some other event), then a search through the ordered list of possible rooms is conducted. If a suitable room is found, then that room can be written back to the genetic representation.

2.2 Local Search Results

There has been some debate as to the need for local search (memetics) in an evolutionary approach to timetabling. Local search can be computationally very expensive and it could be argued that the advantages of local search are outweighed by the disadvantage of processing far fewer chromosomes in the same time. Some methods do not use local search e.g. [2], some use local search throughout e.g. [1], [3], others use local search to seed an initial population e.g. [4].

In order to explore this issue, experiments were conducted using different amounts of local search. The data for these experiments was that for the Computer Studies Department at Napier University. In one experiment local search was always used, in another in was only used to seed the initial population, and in other experiments local search was used when evaluating a randomly selected percentage of chromosomes. Each experiment was given the same elapsed time to run, and so experiments using less local search were able to perform a significantly larger number of evaluations. The results are given in Figure 3 and show the weighted sum of penalty points over 12 competing objectives, averaged over 25 runs. These results show that significant improvement can be made to the algorithm by including local search not just at the population seeding stage, but throughout the whole algorithm.

Penalty Points After Five Minutes

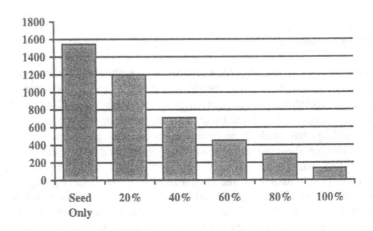

Fig. 3. Effect of varying the amount of local search

Some systems, such as that described in [5], use a method known as delta evaluation. This method cuts the processing necessary for timetable evaluation by re-evaluating only the parts of the timetable that are changed through recombination or mutation. Using local search throughout the evolution makes it impossible to efficiently implement delta evaluation. While the authors expect that delta evaluation cannot compensate for the lack of local search, it should be noted that experiments have not been conducted to examine this.

2.3 Lamarckism

If local search is used then the results of this can be written back into the chromosome giving Lamarckian evolution. There is some debate as to the relative benefits of Lamarckian evolution compared to simply making use of the Baldwin Effect (using local search but not writing the results back) [6].

Experiments were conducted, in a similar fashion to those above, to measure the effect of varying the percentages of chromosomes that have the results of local search written back to them. It should be noted that even in the experiments where no direct writeback occurs, there is still a Lamarckian effect, since the directed and targeted mutation operators have a Lamarckian nature. The results given in Figure 4 show that for this real world problem Lamarckism has a significant advantage over use of the Baldwin Effect. As soon as some writing back is done there is a big improvement. Increasing the amount of writeback does not have a large effect because if there is some writeback in the system then each chromosome will eventually have its local search results written back even if it has to wait some generations for this.

The main reason that writing back is important is that it decreases epistasis in the chromosomes. The local search considers each of the events in turn and tries to place the event in the timeslot and room specified by the chromosome. If the timeslot and room are not available then others are tried. Without writeback, the timeslot and room used by an event at the last evaluation are not necessarily the ones specified in the chromosome. This means that choosing the same timeslot and room is dependent on the slots and rooms tried earlier being unusable - i.e. the resources they need have already been used by some event considered earlier. So the timeslot and room for an event is dependent on the slots used by previously considered events. If this event gets a different slot then so might others considered afterwards. Writing back the results of local search avoids this, so decreasing epistasis in the chromosomes, making then less brittle.

Penalty Points After Five Minutes

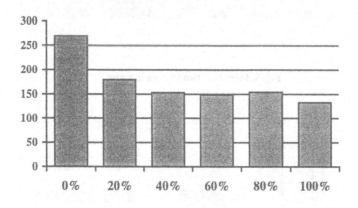

Fig. 4. Effect of varying the amount of writeback

3. Evaluation Criteria and Results

The Napier University Science Faculty, was timetabled using 12 competing objectives in addition to the objective of placing all events. The results obtained from a typical run are shown in Figures 5-16. Results show penalty points for each evaluation criterion are given for run time of 10 minutes, 1 hour and 12 hours on a Pentium Pro 200 computer system, along with the figures for the manually produced timetable.

It can be seen from the results that the pilot study was successful. The system produces feasible timetables in a few seconds. If the system is left to run overnight then timetables can be produced that are better than manually produced timetables in all measured criteria.

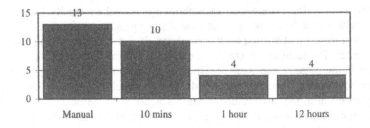

Fig. 5. Lecturers without a teaching free day.

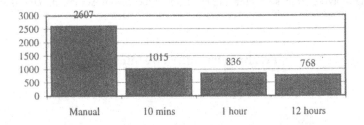

Fig. 6. More than two lectures in a row

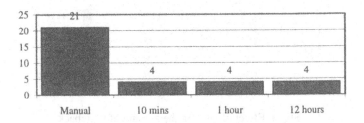

Fig. 7. More than three hours class contact in a row

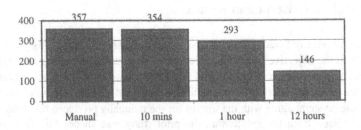

Fig. 8. Gaps of more than three hours in a student's day.

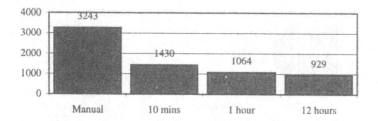

Fig. 9. Wednesday Afternoon Classes.

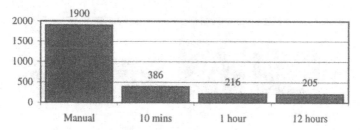

Fig. 10. Five o'clock classes.

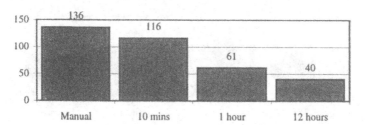

Fig. 11. Single classes on a student's day.

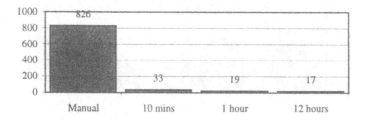

Fig. 12. Site changes during the day.

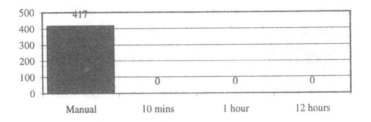

Fig. 13. Instantaneous site changes during the day.

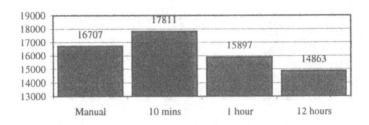

Fig. 14. Location changes within a site.

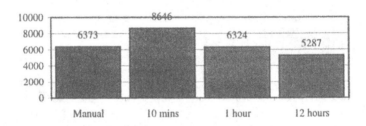

Fig. 15. Room changes within a location.

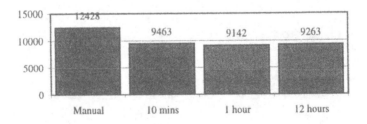

Fig. 16. Seat wastage.

4. Conclusions

We have defined two concepts, *features* and *containers* and shown how they can be used to help define constraints in timetabling problems. We have shown that the policy of using a memetic (local search) algorithm with Lamarckian evolution work well for this problem.

A system has been produced which is easy to use, quickly produces feasible timetables, and which given an overnight run can be produce timetables that are better than those manually produced in all measured criteria.

References

[1] Paechter B., Cumming, A., Norman, M.G. and Luchian, H., "Extensions to a Memetic Timetabling System", in Practice and Theory of Automated Timetabling, Eds., Burke, E. and Ross, P., Springer-Verlag Lecture Notes in Computer Science 1153, Berlin 1996.

[2] Corne, D., Ross, P. and Fang, H, "Fast Practical Evolutionary Timetabling", in Evolutionary Computing, Ed. Fogarty, T, Springer-Verlag Lecture Notes in Computer Science 865, Berlin 1994.

[3] Burke, E. Newall, J.P., and Weare, R.F., "A Memetic Algorithm for University Exam Timetabling", in Practice and Theory of Automated Timetabling, Eds., Burke, E. and Ross, P., Springer-Verlag Lecture Notes in Computer Science 1153, Berlin 1996.

[4] Corne, D and Ross, P., "Peckish Initialisation Strategies for Evolutionary Timetabling", in Practice and Theory of Automated Timetabling, Eds., Burke, E. and Ross, P., Springer-Verlag Lecture Notes in Computer Science 1153, Berlin 1996.

[5] Ross, P., Corne, D. and Fang, H., "Improving Evolutionary Timetabling with Delta Evaluation and Directed Mutation" in Parallel Problem Solving from Nature ii, Ed. Davidor, Y., Springer-Verlag, Berlin 1994.

[6] Turney, P, Whitley, D. and Anderson, R. "Evolution Learning and Instinct: 100 Years of the Baldwin Effect" in Evolutionary Computation, Volume4, Number 3, MIT Press 1996.

Constrained Based Methods

A Constraint-Based Approach for Examination Timetabling Using Local Repair Techniques

Philippe David

École des Mines de Nantes
4 rue Alfred Kastler, La Chantrerie, BP 20722
44307 Nantes Cedex 3 – France
pdavid@emn.fr

Abstract. We present in this paper an algorithm based upon the Constraint Satisfaction Problem model, used at the "École des Mines de Nantes" to generate examination timetables. A strong constraint is that the computing time must be less than 1 minute. This led us to develop an incomplete algorithm, using local repair techniques, instead of an exhaustive search method. The program has been validated on fifty "hand-made" problems, and has succesfully solved the thirteen "real" problems.

1 Introduction

Every year, the "École des Mines de Nantes", a French school of engineering, organizes an entrance examination to select its students. After a first step (written examination), there remain about 1500 candidates. These candidates are then called for a second step consisting of a sequence of oral examinations. Every day, a number of candidates have to take several oral exams. Section 2 describes the complete problem. One important point is that the examination timetable must be generated "on-line": the available time is less than 1 minute. This is obviously not enough for a hand-made timetable. The person in charge of the entrance examination therefore asked us for a program.

Timetabling (both in general and in our particular instance) is an NP-hard problem: there is consequently no guarantee that any complete method can either find a solution or prove there is no solution within 1 minute.

Many approaches exist to address examination timetabling [1, 2]. We chose to represent our problem as a Constraint Satisfaction Problem (CSP) [6]. But we gave up the exhaustive search usually used; instead, we implemented an incomplete assignment algorithm, based upon local repair techniques. Both incomplete methods [5] and local repair techniques [4] have proven their efficiency on certain classes of problems, such as graph coloring or the n-queens problem; also, [4] succesfully solve the problem of scheduling astronomical observations on the Hubble Space Telescope. But as far as we know, very few works include constraint-based approaches associated with repair techniques to generate timetables ; amongst these works, we can cite [7], whose relations with our works will be discussed in section 6.

Of course, an incomplete method can miss solutions. To minimize this risk, the program can be run a number of times: moreover, if no solution is found, new runs are processed, with some constraints *relaxed*.[1]

The program was first validated on fifty "hand-made" problems (specially created to test the program), and finally succeeded to generate all thirteen "real" examination timetable the school asked for.

This paper is organized as follows: the problem we have to solve is described in section 2; section 3 presents Constraint Satisfaction Problems (CSPs) and a formulation of the problem as a CSP. Section 4 explains why we use an *incomplete* algorithm, and presents the algorithm together with its time complexity. Finally, experimental results are presented in section 5.

2 Presentation of the Problem

For each subject, there are several different examiners. The examiners are generally high school teachers.

For a given day, a number of candidates (say n_c) have to take the complete examination, composed of the four following subjects:

1. mathematics (n_m examiners for the considered day),
2. physics and chemistry (n_p examiners),
3. foreign language: either English (n_e examiners), German (n_g examiners) or Spanish (n_s examiners),
4. discussion about a text (n_d examiners).

The total number of examiners (for a given day) will be denoted n_{exa}. The day is composed of p consecutive periods (a period is the time an examination lasts); in our case, there are $p = 15$ periods numbered #1, #2, ... , #15.

2.1 Objective

The aim is to provide a schedule of the examinations for the day: for every subject, assign every candidate to one examiner during one period, so that a number of constraints (presented in section 2.2) are satisfied.

2.2 The Constraints

There are two types of constraints: *required* constraints which must necessarily be satisfied, and *secondary* constraints which should preferably be satisfied, but can be violated.

---

[1] A constraint is called *relaxed* if it can be violated; in other words, relaxing a constraint amounts to removing it from the set of constraints.

Required Constraints

1. every candidate must take an exam in every subject,
2. each candidate takes an exam in the language he chooses (among those proposed) with an examiner of *this* language,
3. for a given period, only one candidate is assigned to any one examiner,
4. any candidate can be assigned to only one examiner during the same period,
5. the candidates are given some time to prepare each exam: they must have at least one free period before each exam; in other words, they cannot take two exams in adjacent periods.

Secondary Constraints

They are presented from the most to the least important:

6. no candidate should meet an examiner who comes from the same school,
7. if the number of candidates is less than the maximum number of examinations, the free periods should be the last ones (#15, #14,...),
8. the examiners should not have free periods during the day,
9. every examiner should have roughly the same number of candidates.

Hereafter, those constraints will be referred to as C_1, ... , C_9. Notice that C_7 and C_8 are two different ways of formulating the same requirement: hereafter, we will call this constraint $C_{7,8}$.

The schedule must be computed every morning with the actual list of candidates and examiners. An important consequence is that *the computing time must be less than 1 minute*. We will see in section 4 the influence this requirement had on the method we chose.

2.3 The Data

The Examiners There are generally 7 examiners for each subject (the 3 *foreign languages* are considered *together* as one subject: $n_e + n_g + n_s = 7$); in most cases, each of them is present for the whole day (the 15 periods). But it may happen that instead of 1 examiner being present during 15 periods, there are 2 examiners: the first one during the first 7 periods (the morning) and the second one during the last 8 periods (the afternoon). The first examiner may also be present during periods #1, #2 #14, #15 and the second one during periods #3 to #13: all combinations are possible; there may even be 15 different examiners. The total number of possible examination periods must quite simply be no less than the number of candidates.

The Candidates During a complete day, at most 105 candidates should take the examination, where $105 = 7 \times 15$ (number of examiners $\times$ number of periods for each examiner). The candidates, as well as the examiners, come from schools in different points of France. It may therefore happen that some candidates come from the same school as some examiners.

3 Representation of the Problem

We represented and solved this problem with a constraint-based approach. We first present the CSP (Constraint Satisfaction Problem) model, and we will then show how we represent our problem as a CSP. In section 4, we will present how and why we adapt the techniques usually implemented to solve CSPs. We will in particular explain why we did not use a "commercial" constraint system.

3.1 Presentation of the Model

A CSP can be defined as a finite set $\mathcal{X}$ of n variables $X_1, X_2, ..., X_n$; the set $\mathcal{D}$ of their respective domains $domain(X_1), domain(X_2), ..., domain(X_n), (domain(X_i)$ is the set of the possible values for X_i); and the set $\mathcal{C}$ of the constraints specifying which combinations of values are allowed.

Solving a CSP consists in assigning a value to every variable so that all constraints are satisfied. The usual method is based upon an exhaustive enumeration algorithm (*e.g. backtrack*), associated with domain reduction techniques [6, 3].

3.2 The Problem as a CSP

We associate a variable to every pair (examiner, period); for instance, if examiner m_1 is present from period #8 to period #15, then 8 variables are considered: $X^8_{m_1}, ..., X^{15}_{m_1}$. The domains are the sets of possible candidates for these pairs (initially, all the candidates). We will see in the following sections how the domains are precisely defined (and reduced), and how the constraints are represented.

Notations:

- Hereafter in this paper, *the domains of an examiner* will represent the set of the domains of all variables related to this examiner.
  Example: the domains of m_1 are $domain(X^8_{m_1}), ..., domain(X^{15}_{m_1})$.
- A candidate c will be called *unavailable* for an examiner for a given period if c is already assigned for this period or an adjacent one (the immediately previous or immediately next period).

4 The Assignment Algorithm

The problem we have to solve is clearly NP-hard. It is therefore impossible to guarantee that a solution can be found (or that it can be proven that no solution exists) within 1 minute, which is an imperative request. After a careful analysis of the actual data, we decided to implement an *incomplete* algorithm. Although there were some theoretical possibilities of missing solutions, this choice seemed to offer the best compromise between *computing time* and *quality of the solution*. We did not implement our program with a commercial constraint system for two main reasons:

1. Commercial constraint systems are based upon a complete enumeration method, which conflicts with our choice of an incomplete method (remember the available time is less than 1 minute), and
2. our problem requires constraint relaxations, which are not implemented in current commercial systems.

4.1 General Principle

The algorithm we used to solve the problem proceeds in two steps:

1. a first step (called *pre-assignment*) creates the domains: every examiner is associated to a list of candidates which satisfy certain constraints (see section 4.2);
2. a second step (the *final assignment*), assigns the candidates to a period, for the examiner whom they are pre-assigned to.

Obviously, if the first step fails, the second step is not processed.

4.2 The Pre-assignment Step

A first possibility could have been to create the same domain for all variables, with all candidates. The domains would have then been reduced by different constraints (*e.g* candidates removed from domains of examiners coming from the same school).

We chose another possibility: some constraints are taken into account during the creation of the domains.

1. First, the number of candidates is computed for each examiner to satisfy C_9.
2. Second, for each subject, every candidate is inserted into all the domains of an examiner, so that constraints C_2 (for the language examiners) and C_6 are satisfied, until all examiners have the number of candidates computed in step 1. All candidates are then pre-assigned (each one to 4 examiners – one for each subject).

Complexity

The pre-assignment step is straightforward: the complexity is therefore $O(n_c)$.[2]

---

[2] One could say that each candidate is assigned to *every domain* of an examiner (at most p domains), which leads to an $O(n_c p)$ complexity. But all initial domains of the examiners are equal to each other. Only one domain can therefore be constructed; the domains of all examiners will be duplicated in a later process, ($O(n_{exa} p)$ duplications).

4.3 The Final Assignment Step

While the pre-assignment step consists in "filling" the domains, one can say that the final assignment step empties them.

The pre-assignment process together with the way the final assignment is achieved ensure that all required constraints are satisfied:

- when a candidate is assigned to an examiner for a period, then this candidate is removed from all other variables of this examiner (the other periods): this ensures that every candidate will only take one exam in each subject (remember that the pre-assignment creates domains in such a way that a candidate belongs to the domains of only one examiner for each subject);
- the assignment step succeeds only when all candidates have been assigned. These two points ensure that constraint C_1 is satisfied.
- Constraint C_2 is satisfied by definition of the pre-assignment step.
- Constraint C_3 is clearly satisfied by definition of the model: any variable is assigned one and only one value.
- Constraints C_4 and C_5: when a candidate is assigned to an examiner for a given period, this candidate is marked *unavailable* in all domains containing him for this period (C_4), the previous one and the next one (C_5).

This can be achieved by removing the candidate from the related domains.

Two cases may happen:

1. some domains remain empty and not all candidates are assigned: the assignment step fails (this is the case if all remaining candidates for a given examiner and a given period are unavailable). We can however try to "repair" the assignment, using local changes: this is achieved by several **local repair** procedures, presented in section 4.4.
2. all candidates are assigned: the assignment step succeeds. Although some variables may be "unassigned", for instance if there are less candidates than possible periods. Those variables (which correspond to periods during which the examiner is free) are then assigned a special value.

Some constraints are so far not taken into account: we now explain how they are handled.

- Constraint $C_{7,8}$ is satisfied by choosing a "good" assignment ordering: the variables are assigned "period by period": all variables corresponding to period #1 are assigned, then period #2, and so on.

We also try to minimize the risks of future failures (empty domains – see above). We chose two heuristics to limit these risks:

1. for the current period, we first choose the examiner with the smallest domain (least number of available candidates); this is a classical heuristic in CSP ("first-fail" ordering).
2. we then choose the least restrictive candidate for the future choice of the other examiners; in other words:

- for every available candidate (for the current examiner and the current period):
  - for every other examiner this candidate is pre-affected to, the number of other available candidates is computed;
  - the smallest number is associated to this candidate;
- the candidate with the largest number is chosen.

We can now sketch the global assignment algorithm: see figure 1.

Step 1: pre-assignment

```
1   for every subject, assign each candidate to the domains
    of an examiner, as described in section 4.2
```

Step2: final assignment

```
2   for all periods j from 1 to 15
3     while all variables of the period are not assigned
4       if all candidates are assigned, return success
5       let Xⁱʲ be the examiner with the smallest domain // for period j
6       if domain(Xⁱʲ) is not empty:
7         extract the less restrictive candidate from domain(Xⁱʲ);
          let this candidate be denoted cand
8         assign cand to Xⁱʲ
9         go to 3 // continue: try next candidate
10      if domain(Xⁱʲ) is empty // try to repair the assignment
11        if find-in-past succeeds
12          go to 3
13        else if find-with-other succeeds
14          go to 3
15        else if find-with-other-past succeeds
16          go to 3
17        else if swap-examiner-and-check succeeds
18          go to 3
19        else leave Xⁱʲ unassigned // all repair trials failed
20        end if
21      end if
22    end while
23  end for
24  if all candidates are not assigned, return failure
```

Fig. 1. General view of the assignment algorithm

A few comments about line 19: if all successive repair trials fail, the current examiner is left unassigned for the current period. The assignment will possibly still succeed if the examiner has more possible periods than candidates to examine. In this case, the examiner will have this period (#j) free in his timetable. That corresponds to a relaxation of constraint $C_{7,8}$.

Another remark: as we said in section 4, this algorithm is *incomplete*. If some constraints empty a domain, there is no backtrack. Some previous assignments can possibly be changed (by the repair procedures), but these possible changes are local: there is no exhaustive search. This is a key point in our algorithm, and makes the problem tractable, but it also means that a solution may be missed. We will see in section 5 that it was (fortunately ?) not the case during the actual examinations.

4.4 The Local Repairs

There are five different repair procedures, which are successively called upon if the previous one fails. We now present them in the order they are called (see figure 1). They all follow the same principle: find a candidate to assign to the current examiner, for the current period.

Find-in-past No candidate not yet assigned (say, to examiner **exa**) is available. A candidate therefore has to be found amongst those already assigned (to **exa**). To do so, we try to replace a candidate assigned previously to a former period (say **past-period**). To free him, we look for a pair (**unassigned candidate, assigned candidate**) in which **unassigned candidate** is available for **past-period** and **assigned candidate** is available for **current-period**. **Unassigned candidate** is then assigned to **past-period** instead of **assigned candidate**, who can now be assigned to **current-period**.

Complexity
Let a be the number of candidates already assigned to the current examiner (the "past candidates"), and r the number of remaining (unavailable) candidates; let c be the total number of candidates the current examiner has to examine during the day: $c = a + r$. For each unavailable candidate, every past candidate is possibly checked. The total number of checks is at most $a \times r$, whose greatest value is $c^2/4$ when $a = r = c/2$. As $c \leq p$ (remember p is the number of periods), the worst case complexity of procedure **find-in-past** is therefore $O(p^2)$.

Example of processing of **find-in-past** *(Figure 2):*
We try to assign a candidate to examiner j for period **4**; 2 candidates remain to be assigned, both of them are unavailable:

- candidate **4** was assigned to examiner j-i for period 3;
- candidate **5** was assigned to examiner j+k for period 4.

We therefore try to free a candidate assigned to this examiner (j) for a former period:

- We first try candidate 4 (denoted c4):
  - c4 is not available for period #1 (he is already assigned to examiner j+k for period #1);
  - c4 is not available for period #2 (he is already assigned to examiner j+k for period #1, and to examiner j-i for period #3);

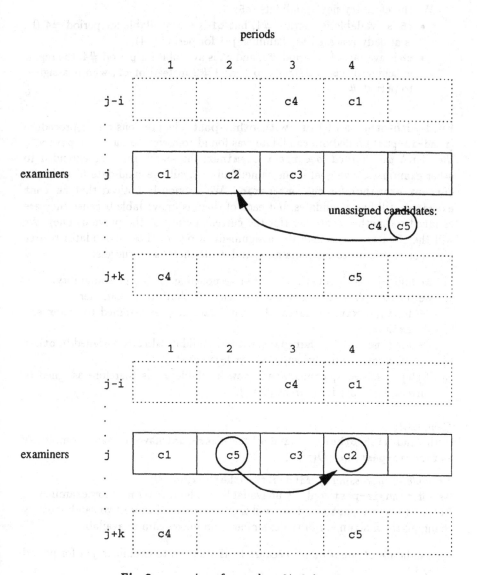

Fig. 2. processing of procedure f ind-in-past

- c4 is not available for period #3 (he is already assigned to examiner j-i for period #3);

Candidate c4 therefore cannot be assigned to any period before #4.

- We therefore try next candidate, c5:
  - c5 is available for period #1, but c1 is not available for period #4 (he is already assigned to examiner j-i for period #4);
  - c5 is available for period #2, and c2 is available for period #4: the repair is achieved: c5 is assigned to period #2 instead of c2, who is assigned to period #4.

Find-with-other and Find-with-other-past The previous repair procedure (find-in-past) failed: no candidate was found to assign to current period *if the search was limited to* current examiner; the search must be extended to other examiners. The goal is the same: find an available candidate to assign to current examiner for current period. We previously noticed that current examiner still has candidates, but each of them is unavailable because they are assigned to another examiner (for the current period or the previous one). We will therefore try to change the assignment of one of those candidates to free him. For each candidate remaining, and while there is no success:

1. we find who is the other examiner responsible for the unavailability;
2. we try to replace the candidate assigned to that other examiner:
   - for find-with-other, with a candidate not yet assigned to other examiner,
   - for find-with-other-past, with a candidate already assigned to other examiner (same principle as find-in-past);
3. if step 2 succeeds, candidate is now available; he is therefore assigned to current examiner for current period.

Complexity
Both find-with-other and find-with-other-past have the same complexity as find-in-past, *i.e.*, $O(p^2)$.

Example of processing of find-with-other *(Figure 3):*
As for find-in-past, only 2 unavailable candidates remain for examiner j. But, as find-in-past failed, we will try to free one of those two candidates by changing the assignment of the examiner who makes him unavailable.

- candidate c4, unavailable because he is assigned to examiner j-i for period #3:
  - We look for a candidate to replace c4 (amongst those who are not assigned yet to examiner j-i);
  - candidate c7 is available for period #3;
  - candidate c7 is therefore assigned to examiner j-i for period #3, instead of c4, who is now available for examiner j and period #4 (note that c4 is re-inserted into the list of unassigned candidates for examiner j-i).

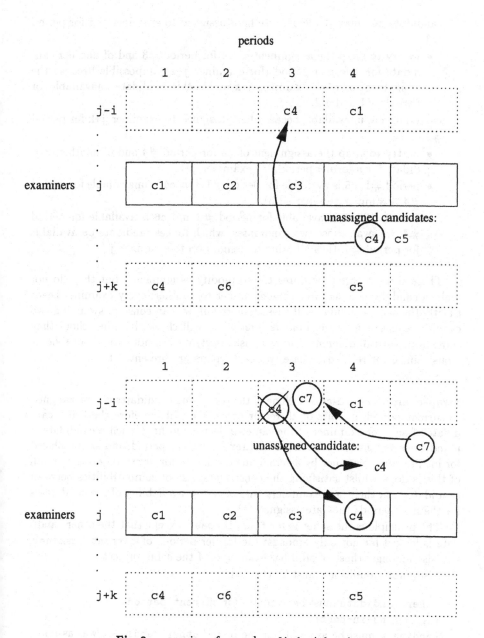

Fig. 3. processing of procedure find-with-other

Example of processing of find-with-other-past *(Figure 4):*
Same initial conditions as find-with-other.

- candidate c4, unavailable because he is assigned to examiner j-i for period #3:
  - we try to swap the assignment of c4 for period #3 and of another candidate for a former period (for examiner j-i): impossible because the assignment of c4 to examiner j+k makes this candidate unavailable for periods #1 and #2.
- candidate c5, unavailable because he is assigned to examiner j+k for period #4:
  - we try to swap the assignment of c5 for period #4 and of another candidate for a former period, for examiner j+k:
  - period #1: c5 is available for period #1, but c4 is unavailable for period #4 (examiner j-i, period 3);
  - period #2: c5 is available for period #2 and c6 is available for period #4: we swap those two candidates, which makes candidate c5 available for period #4, thus allowing to assign him to examiner j.

These three repair procedures try to modify assignments, but they do not make a candidate change from one examiner to another: every examiner keeps exactly the same candidates. If these repairs fail, we can consider swapping two candidates between two examiners; firstly, we will check that the school they come from are still different, and as a last resort, we will not care about schools (constraint C_6 is relaxed). These two new repairs are presented below.

Swap-examiner-and-check None of the remaining candidates can be assigned to current period, in spite of all repair trials. We will therefore check if a candidate of another examiner of the same subject could fit. If such a candidate is found, he is assigned to **current examiner** for current period, and he is replaced for his previous examiner by a remaining candidate for **current examiner** (all of this is done whilst satisfying the constraints: school compatibilities between candidates and their new examiners, and of course availability of the candidates for the new periods they are assigned to).

The principle is the same as for find-in-past, except that the other candidate is looked for not only amongst the former periods of **current examiner**, but also amongst those of all other examiners of the same subject.

Also, more conditions must be checked:

- **other candidate** must be available for **current period** (as for find-in-past);
- **candidate** must be available for the period **other candidate** was assigned to (also, as for find-in-past);
  but also:
- **other candidate** must not come from the same school as **current examiner**;

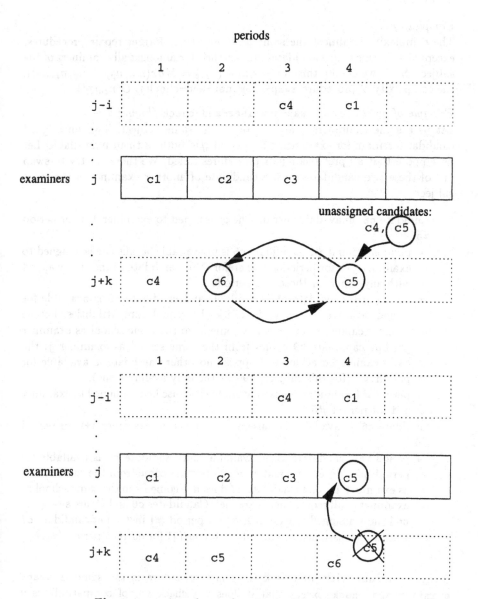

Fig. 4. processing of procedure find-with-other-past

– **candidate** must not come from the same school as the former examiner of **other candidate** (since he will be assigned to this examiner in case of a swap).

Complexity

The complexity is almost the same as those of the former repair procedures, except that in this case, candidates are searched for amongst all examiners of the subject. So, if we denote this number n_{sub} ($n_{sub} = Max(n_m, n_p, n_e, n_g, n_s, n_d)$), the complexity of procedure **swap-examiner-and-check** is $O(n_{sub}p^2)$.

Example of processing of **swap-examiner-and-check** *(Figure 5):*

Let us assume examiners j and j' teach the same subject. Two unassigned candidates remain for examiner j for period #5, both of them unavailable. Let us suppose that all previous repair procedures failed. We therefore try to swap one of these two candidates with a candidate of another examiner of the same subject:

– candidate c5, unavailable because he is assigned to examiner j-i for period #5:
  - periods #1 and #2: candidate c5 is unavailable because he is assigned to examiner j+k for period #1: therefore this candidate cannot be swapped with any other for those two periods.
  - period #3: candidate c5 is available and candidate c9 is available for period #5. We now have to check the school compatibilities. Let us suppose candidate c5 does not come from the same school as examiner j', but candidate c9 comes from the same school as examiner j: the swap cannot be achieved. Suppose no other candidate is available for period #3 (for this subject, c9 was the only available one).
  - period #4: candidate c5 is unavailable because he is assigned to examiner j-i for period #5.
– candidate c6, unavailable because he is assigned to examiner j+k for period #4:
  - period #1: candidate c6 is available and candidate c7 is available for period #5. Moreover, candidate c6 does not come from the same school as examiner j', and candidate c7 does not come from the same school as examiner j: all requirements are met. Candidates c6 and c7 are swapped: c6 is now assigned to examiner j' for period #1 instead of candidate c7 who is now assigned to current examiner (j) for current period (#5).

Swap-examiner-without-check This last procedure is the same as **swap-examiner-and-check**, except that it does not check school compatibilities if candidates swap from one examiner to another (*i.e*, constraint C_6 is relaxed).

Notice that constraint C_9 can be relaxed by procedures **swap-examiner-without-check** or **swap-examiner-and-check** by "giving" a candidate from one examiner ei to another examiner ej which is achieved by swapping this candidate with a free period of ej.

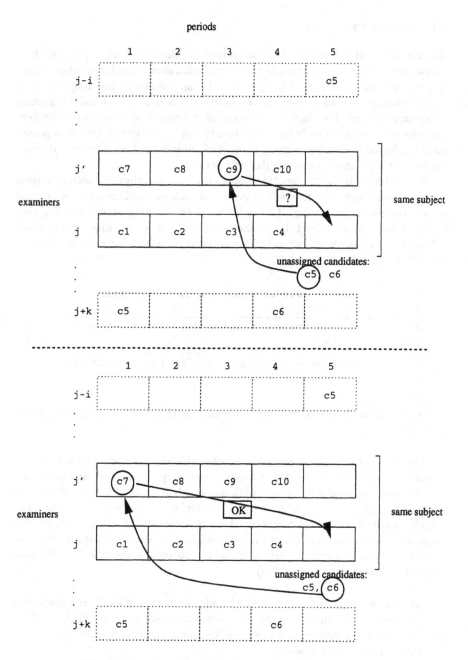

Fig. 5. processing of procedure swap-examiner-and-check

4.5   Failure Processing

The algorithm we use to solve this assignment problem is not a complete method. Consequently it is possible for the algorithm to fail even though a solution exists. A possible reason for such a failure could be an "unlucky" pre-assignment, in that it cannot lead to a solution. To limit such a risk, the global assignment processing presented in figure 1 is repeated a number of times, with different random pre-assignments (which still satisfy the constraints); this is repeated until a solution is found or a given number of trials is reached. During this cycle, constraints $C_{7,8}$ and C_9 are possibly relaxed.

If no solution is found after this first cycle of trials, a slightly "softer" version (certain constraints are relaxed) is processed: school conflicts are not checked during the repair processing. This is achieved by inserting an extra repair procedure (`swap-examiner-without-check`) after the first four procedures, which all check school conflicts; we only have to insert two lines between line 18 and line 19 of figure 1.

```
...
17        else if swap-examiner-and-check succeeds
18            go to 3
18.2      else if swap-examiner-without-check succeeds
18.3          go to 3
19        else leave X_i^j unassigned  // all repair trials failed
20        end if
...
```

A second cycle of trials is then processed, with the added relaxation of C_6 possible.

Finally, if there is still no solution after this second cycle of trials, a still softer version is processed. The pre-assignment step is removed: any candidate can be assigned to any examiner, with no no attention paid to the school they come from. The local repair procedures are replaced by adapted versions (*i.e.*, which do not check school conflicts). After a "pre-solution" (the assignment is complete, but some candidates may be assigned to examiners from the same school) is found, it will be processed to try to reduce the number of school conflicts (by swapping candidates).

As a last resort, some periods could be added after the #15. After a discussion with the person in charge of the entrance examination, we decided to add 2 periods: it seemed sufficient to solve possible failures, and it did not increase the examiners workload excessively.

4.6   Overall Complexity

Complexity of step 1 is $O(n_c)$, or $O(n_c + n_{exa}p)$ if the duplication stage is included. In step 2 each candidate is either assigned to a variable (examiner,period)

– $O(1)$ –, or processed by repair procedures – $O(n_{sub}p^2)$ –. Step 2 is therefore $O(n_c n_{sub}p^2)$. But $n_{sub}p$ roughly equals n_c. Step 2 complexity can thus be simplified to $O(n_c^2 p)$, which is therefore the complexity of the whole assignment algorithm.

5 Validation and Results

We paid a particular attention to the validation of the program, in particular because of the following two characteristics:

1. we implemented an incomplete algorithm, which implies a theoretical risk of missing solutions;
2. the program was to be run "on-line" (candidates and examiners are waiting for their timetabling); it is not possible to tell them
   "Sorry, but we cannot give you a timetable! Come back later..."

In order to achieve the validation stage with sufficient guarantees, we ran the program with fifty test problems, specially designed to simulate all cases which could potentially give rise to difficulties (or, at least, all those we could think at): unavoidable school conflicts, many examiners only available for a few periods (restriction of repair possibilities), more candidates than periods (*i.e* only one Spanish-speaking examiner, and 16 (or even 20) Spanish-speaking candidates)...

This last case obviously failed: no solution exists. All other tests could generate a solution; most of them satisfied all the constraints. We carefully examined the timetables with some relaxed constraints, and we could not find a better solution (*i.e.*, with less (or less important) relaxed constraints).

All those tests were achieved with run times between 0.2 and 7.5 seconds on a SUN SPARC 5 with 32 Mbytes of memory. We found this experiment conclusive. Fortunately, the real assignment problems confirmed that we were right.

On each of the thirteen days the program was run with actual examiners and candidates, a timetable was generated, with no relaxed constraint. Moreover, all run times were either 0.2 or 0.3 seconds. Although we did not compute all solutions, we suppose that such good results are due to the nature of the data; there are probably many solutions, in particular due to symmetries.

6 Conclusion

We have presented in this paper the algorithm we designed and implemented to generate an examination timetable for the "École des Mines de Nantes". We represented the assignment problem as a Constraint Satisfaction Problem, but we did not solve it with a usual method based upon an exhaustive search. Instead, we implemented an incomplete algorithm, together with some *ad hoc* techniques:

– heuristics to limit the risk of constraint violations;
– local repair procedures to reduce the number of constraint violations.

These techniques can be compared to those proposed in [7]; their method differs from "classical" hill-climbing or repair techniques in that theirs do not randomly generate initial assignments; they use an arc-consistency algorithm to generate initial assignments, which they then refine with a hill climbing algorithm. The generation step is therefore more time consuming, but the quality of initial assignments is much higher.

On the other hand, *exhaustive search* algorithms are much more time consuming (they need exponential time), but they guarantee a solution is found (if it exists) or *the best* solution is found (for optimization problems). Obviously, those algorithms are not followed by an iterative improvement step such as hill-climbing or repair techniques.

Our method can be classified between those exhaustive search algorithms and [7]: we repair the solution *during* its generation. We go further in the compromise between computing time and quality of the assignment.

In our case, where computing time is critical, there is no time to try to improve the initial solution iteratively. This explains why we chose to spend more time in the generation step to avoid this final improvement phase.

We think applying such methods to other timetabling problems could give interesting results, in terms of computing time as well as in terms of quality of solution: the number and the level of repair procedures can be adapted to every specific situation.

To conclude, these techniques were particularly well-suited for our application: the program succesfully generated all the timetables we were asked for and it will be distributed to other "Écoles des Mines" schools in France.

References

1. Michael W. Carter. A survey of practical applications of examination timetabling algorithms. *Operations Research*, 34(2):193–202, 1986.
2. Michael W. Carter and Gilbert Laporte. Recent Developments in Practical Examination Timetabling. In Edmund Burke and Peter Ross, editors, *Practice and Theory of Automated Timetabling*, volume 1153 of *Lectures Notes in Computer Science*, pages 3–21. Springer-Verlag, 1996.
3. Alan K. Mackworth and Eugene C. Freuder. The complexity of constraint satisfaction revisited. *Artificial Intelligence*, 59:57–62, 1993.
4. Steven Minton, Mark D. Johnston, Andrew B. Philips, and Philip Laird. Minimizing conflicts: a heuristic repair method for constraint satisfaction and scheduling problems. *Artificial Intelligence*, 58(1–3):161–205, December 1992.
5. Bart Selman, Hector Levesque, and David Mitchell. A new method for solving hard satisfiability problems. In *Proceedings of the Ninth National Conference on Artificial Intelligence (AAAI-92)*, pages 440–446, San Jose, CA, 1992.
6. Edward Tsang. *Foundations of constraint satisfaction*. Academic Press, 1993.
7. M. Yoshikawa and K. Kaneko and T. Yamanouchi and M. Watanabe. *A Constraint-Based High School Scheduling System*. *IEEE Expert*, 11(1):63–72, February 1996.

Generating Complete University Timetables by Combining Tabu Search with Constraint Logic

George M. White
Junhan Zhang

School of Information Technology and Engineering,
University of Ottawa,
Ottawa, K1N 6N5,
Canada
white@site.uottawa.ca

Abstract. Several small data sets representing a few university departments were used with both a constraint logic program and a tabu search program to cast a timetable. The constraint logic program used alone produced timetables rather quickly. The tabu search program used alone ultimately produced better solutions but at a much slower rate. The sequential use of a constraint logic program whose output was used to start the tabu search produced the best timetables of all in a time that was much longer than that of the logic program alone but shorter than that of the tabu search used alone.

1. Introduction

The automatic generation of university and college timetables that are able to satisfy institutional requirements is a subject that has been studied for some time by many researchers. A survey of most aspects of this problem has recently been published by Bardadym[1] . It has proved to be a difficult problem because of both its inherent complexity and the large amount of data that is required to test realistic timetables that bear some resemblance to problems encountered in the real world.

Many related problems have been studied and some, notably the examination scheduling problem, have seen much theoretical and practical success[2 ,3] . Some of the solutions to the examination scheduling problem have been commercialized and are in common use in academia.

The timetabling problem, however, has for many years resisted real practical solutions in spite of the amount of labour poured into it. In most, although not all, institutions of higher learning, the timetable is finely crafted by hand, using labour intensive techniques that have been developed over many years. Two books[4 ,5] discuss these manual methods.

It is only in recent years that a renaissance of activity has led to the discovery of techniques capable of generating timetables of acceptable quality. Some of these have entered the commercial market. Research in this area is an on-going activity and improvements are being made continually. This recent progress has increased interest in all aspects of the subject. Bardadym's survey[1] presents a graph showing the enormous rise in the number of papers published on the subject in the past year.

If a single goal could be formulated to describe the aspirations of timetabling research groups, it would be this:

An automatic timetabling system should formulate complete descriptions of which students and which teachers should meet, at what locations, at what times and should accomplish this quickly and cheaply while respecting the traditions of the institution and pleasing most of the people involved most of the time.

It is easier to state the goal than to achieve it. Different institutions have different ideas as to just what constitutes a "good" timetable and a solution considered suitable for one institution might well not be acceptable at another.

In this paper, two successful approaches will be described, the constraint logic approach and the tabu search. Some recent results achieved by these approaches will be compared and the resources required to achieve them will be discussed. A hybrid approach that combines features of constraint logic and tabu search will be presented. It will be demonstrated that this approach can generate timetables of better quality than either constraint logic or tabu separately, although such timetables take much more time to produce than those using only constraint logic. However they take much less time to produce than methods using the tabu search alone.

This approach is similar to one described by Corne and Ross[6] in which an initial solution is found by a 'peckish' timetable construction algorithm, whose output is then used by an evolutionary algorithm or hillclimbing algorithm. The basic idea behind these tandem methods is to speed up the overall timetable construction by using more than one method, each method being applied under circumstances where it performs best.

2. Tabu Search

The tabu search is an heuristic procedure for resolving problems of optimization. It is specifically designed to avoid the trap of local optimality and has found application in a number of domains. An overview of this method has been published by Glover[7] .

Tabu search has three primary features:
• attribute-based memory structures are used to evaluate the various criteria and store historical search information.
• the control mechanism simultaneously uses features which constrain and liberate the search process. These are embedded in the tabu restrictions and the aspiration criteria.
• the memory functions used incorporate both short term and long term attention spans. The short-term memory function aggressively tries to make the best move satisfying certain constraints. These constraints, called the tabu restrictions, are formulated to prevent oscillations or cycles of moves by declaring certain attributes of these moves to be *tabu*. These restrictions permit the method to move beyond the point of local optimality in what is hoped to be an efficient fashion.

A very compact outline of the tabu search algorithm is shown in figure 1, based on a description presented by de Werra[8] .

The method of tabu search can be applied to solving the timetable problem by formulating the requirements of the institution involved and setting a penalty on each instance where the requirements are not satisfied, *e.g.* where a course's teacher is required to teach during a time that is inconvenient, or when a course is scheduled into a room that is located somewhere inconvenient for the students. Such infelicitous course scheduling instances each have a penalty and when these penalties are added up over the entire timetable, a global objective function value is obtained. The tabu search algorithm tries to minimize the value of this objective function by moving the course instances around, trying different times and locations. The goal is to find those times and places that minimize the value of the objective function.

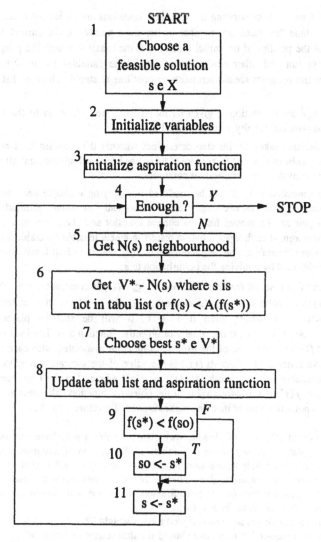

Figure 1

The tabu search algorithm as presented in figure 1 incorporates the timetable requirements in the following way:

Step 1: An initial timetable $s \in X$ is chosen by some method. X is a set of *feasible* solutions where feasible means that all primary constraints are satisfied. One member of this set s is identified either by choosing a random member of the set or by using some heuristic method to generate one. In the work reported here, two methods of initial solution generation were used. The first method randomly generated a trial timetable directly from input data files and relaxed the requirement that the members of X be feasible. The second method used the solution from a constraint logic based implementation, μCTS, described previously[9].

Step 2: Certain variables controlling the stopping conditions are initialized here. These variables include the maximum number of iterations permitted, the current number, the value of the penalty of the initial timetable s, the maximum time the program is permitted to run and other similar quantities. These variables are used to determine whether the program should terminate or continue at step 4. The tabu list T is set to empty.

Step 3: The aspiration function is given its initial value here. It is set to the value of the smallest possible penalty, currently equal to 1.

Step 4: The decision taken in this step determines whether the program halts or not. If the penalty has been reduced to zero, or if there has been no improvement after five iterations, execution halts. Otherwise it continues.

Step 5: Each timetable $s \in X$ can be modified by applying a simple perturbation, e.g. 2 class-teacher pairs can exchange their timeslots and classrooms. Similarly one class-teacher pair can be moved from its current timeslot and classroom to a new timeslot and classroom. If such a perturbation results in a new feasible timetable it is said to be a neighbour timetable. The neighbourhood, $N(s)$, consists of all feasible solutions that can be obtained by applying the perturbation to s.

Step 6: An additional set of feasible solutions $V^* \subseteq N(s)$ is constructed from the elements of $N(s)$ in the following way. An element $s \in N(s)$ is placed in V^* unless s is in the tabu list T. However, even if $s \in T$, it will be still be placed in V^* if $f(s) < A(f(s))$. $f(s)$ is the penalty or objective function associated with timetable s and $A(f(s))$ is the value of the aspiration function associated with each value of the objective function f. Thus $A(f(s))$ is the value of the aspiration function associated with timetable s. In our work, $A(f(s)) = f(s^\circ)$ where s° is the best timetable found so far i.e. $f(s^\circ)$ is the lowest value of the objective function calculated as of yet. Note that s is not a member of its own neighbourhood. Therefore $s \notin V^*$.

$f(s)$ is calculated in the following way. Any timetable s may have zero or more constraint violations of any of the requirements. A violation of a requirement is given a weight and these weights are summed to form the objective function. A requirement that is deemed important is given a heavy weight and one that is not deemed important is given a lighter weight. The requirement violations and weight penalties for not enforcing them are described below:
- course is not taught in a pre-assigned room - weight 23
- teacher is required to teach two classes simultaneously - weight 19
- teacher is required to teach at a time he considers undesirable - weight 13
- required courses of the same program are taught simultaneously - weight 10
- time that course is taught is not in the requested timezone - weight 9
- the room capacity is exceeded - weight 7
- assigned room is not in the requested building - weight 7
- assigned room was in an "excluded" building and nothing was requested - weight 7
- assigned room is not in a "neutral" building and nothing was requested - weight 3
- a 3 hour course is split into two nonconsecutive 1.5 hour blocks - weight 2
- two 1.5 hour sessions of the same course are held on the same day - weight 2
- two sessions of the same course are too close together or too far apart - weight 2
- two sessions of the same course are not held in the same room - weight 1.

Step 7: The penalties $f(s^*)$ associated with each $s^* \in V^*$ are calculated and timetable s^* having the lowest value of the objective function, $f(s^*)$ is chosen. This is done even if s^* is worse than the current timetable.

Step 8: The previous timetable is appended to the tabu list and the aspiration function is updated.

Steps 9 and 10: The new penalty $f(s^*)$ is compared to the lowest penalty found so far $f(s^\circ)$. If $f(s^*) < f(s^\circ)$, then s° is updated to the new timetable. *i.e.* $s^\circ \leftarrow s^*$.

Step 11: In any case, the current timetable is updated: $s \leftarrow s^*$.

Application of the tabu search algorithm to the course scheduling problem has been done by several groups of researchers[10 ,11 ,12] . One of the first applications of the tabu search to the timetabling problem was done by Hertz[10] . He defines a feasible timetable to be an assignment that considers the following:

• courses given at the same time which share a common teacher.

• courses given at the same time which involve common students.

• students or teachers who do not have enough time to move from one building to another.

The objective function involves the number of students and teachers who have conflicts and the amount of time involved in moving from one building to another. An implementation involving 288 courses (10% being preassigned), 143 teachers, 1729 students and 67 classrooms in 15 buildings took 11 seconds to find a solution on a CDC Cyber 170/855.

In Costa's work[11] the objective function is formulated to have 7 components;

• how many teachers are involved simultaneously in two lectures.

• how many classes attend simultaneously two lectures.

• how many times the availability of special rooms is not respected.

• how many lectures are scheduled at a period during which the teaching of a subject is undesirable.

• the number of teachers and classes taking part in two lectures scheduled consecutively during the same day in two distant buildings.

• the degree of compactness of the timetable.

• the sum of the difference in absolute value between the number of classes having a lecture at the first lunch break period and the number of classes having a lecture at the following period. This helps equalize the number of students eating lunch in the two consecutive lunch break periods.

An implementation has been able to produce a timetable for a high school having 780 lectures, 65 teachers, 32 classes, 37 subjects, 12 room types and 50 periods. This required 41 minutes of CPU time on their machine and the result compared well with a manual solution which took about 2 weeks to produce.

The work of Schaerf and Schaerf[12] was done somewhat differently. They tried several different approaches on an implementation running on a Sun SPARC 10. The running time was fixed at 3 hours after which the program was stopped and the result examined. The data was from a school having 10 classes, 20 teachers and 30 periods. They concluded that the tabu search program was superior to their implementation of both a simulated annealing and a randomized descent algorithm. They also found that the tabu search was far more efficient if it were given its initial value from the results of another solution method, such as the solution obtained from their randomized descent program.

3. Constraint Logic

Constraint logic is a form of logic programming in which the variables are subject to constraints. When problems are formulated as constraint logic problems they become known as *Constraint Satisfaction Problems*, (CSP). These are formally defined as follows:[13]

For a finite set of variables, $V = X_1, X_2, \cdots, X_n$, which take their values from their corresponding domains $D_1, D_2, \cdots, D_n$, and a finite set of constraints of the form $c(X_{i_1}, \cdots, X_{i_k})$ among k variables from V representing a subset of the cartesian product $D_{i_1} \times \cdots \times D_{i_k}$ which specifies which values of the variables are compatible with each other, a solution to a CSP is an assignment to all variables which satisfy all the constraints.

To use constraint logic to solve the timetabling problem, the institutional requirements are formulated as constraints on the variables of interest. These variables are some subset of *course, time, room, teacher*. These variables do not have standard meanings; our definitions are found in a previous paper[9].

At the University of Ottawa, the course (more properly, the course section) is assigned to a professor before the timetabling is done. Thus a composite *course-teacher* variable is defined with properties such as (i) the number of students enrolled in that course, (ii) the set of courses that must not conflict with the scheduling of this course and (iii) the teacher's time preferences, as well as several others.

The institutional requirements as formulated into logical constraints are fully described elsewhere[14]. There are two types of constraints. The *primary* constraints must always be satisfied. A feasible timetable satisfies all of them. The *secondary constraints* may be relaxed if it is required to schedule a course such that the primary requirements are satisfied. Otherwise the secondary constraints should be satisfied as well. The constraints chosen are those presently enforced at the University of Ottawa and are outlined below.

3.1. Primary Constraints

1) No two or more required courses of a student program can be scheduled at the same time.

2) No recommended optional course of a student program can be scheduled at the same time as any required course of the same student program.

3) No two or more courses taught by professor can be scheduled at the same time.

4) Day courses must be scheduled during the day and evening courses must be scheduled during the evening.

5) The classrooms assigned to courses must have capacities equal to or larger than the enrollments of the courses scheduled into them.

6) Classrooms must not be scheduled during those times they have been reserved for other uses.

7) Classes cannot be scheduled when their professors are unavailable.

8) Preassigned courses must be scheduled into their preassigned classrooms at their preassigned times. There are three different kinds of preassignment. The first case occurs when a course's teaching time and classroom are all fixed before the scheduling; these are completely preassigned courses. The second case occurs when only the teaching time of the course is fixed; these are called time-preassigned. The final case preassigns a certain classroom to a course.

9) Certain courses must be taught in the same classroom at the same time. This is required when a course has more than one course code, as happens sometimes when a course is taken by students from different departments or faculties.

10) Certain courses must overlap in their teaching times.

11) Certain courses should have exactly the same teaching time but different classrooms. This constraint and the one above can be used to enforce course prerequisites.

12) If a course is held more than once a week, there will always be at least one day between any two lectures of the course.

3.2. Secondary Constraints

The following list of constraints will be imposed if it is possible to do so. Some of these have been added for improving the resource utilization and some are present for improving students' course choice. If it proves impossible to create a schedule with all secondary constraints, they are relaxed in groups in order to decrease the program's run time.

13) Courses should be scheduled into classrooms in the buildings as close as possible to a designated building. The first attempt will be the preferred building. If this is infeasible, then the next and succeeding buildings will be tried until a success is found.

14) Courses should be scheduled into rooms of the specified type. The *type* property of a room has three parameters: class type, furniture type and seating type. Any or all of the types can be specified. The system will first try to find rooms satisfying all three arguments. If they cannot be found, rooms which satisfy room type and furniture will be sought. If this fails, rooms satisfying only room type will be sought. If this fails, the constraint is abandoned.

15) Courses should be scheduled into classrooms whose size is large enough (but not too large) to contain them. The classroom should be in the smallest size range possible. Note that this is not the same as constraint 5.

16) Recommended optional courses of a student program should be scheduled in such a way that there are the fewest time conflicts among them.

17) Courses should be scheduled into the time zone specified by the department or the professor. The time slots available for day courses are partitioned into three time zones, morning, noon and afternoon. While evening courses are always scheduled into the evening time zone, a day course should be scheduled into the prescribed day zone if possible.

18) Friday afternoon shall be used only after all other scheduling possibilities are exhausted.

19) A professor should have at least one time slot free between any two teaching time slots.

Some of the constraints are at least partially in conflict with one another. This is particularly true with constraints 13 and 15 which deal with geography and room size. Constraint 15 will attempt to put a class of students into a classroom just large enough to hold them. This leads to efficient use of classrooms but may require that students and professors do a lot of walking between classes to get to those classrooms. Constraint 13 will minimize the amount of transportation (at least for professors) but may lead to an inefficient use of classrooms.

The 19 constraints listed above are written as the conjunction of sub-rules. They were translated into Prolog with a structure described in[9] forming the rules used by the Prolog interpreter as a controlled deduction. These rules are interpreted by a modified Prolog backtracking algorithm based on depth first search. The classical method is not efficient enough for our purposes and modifications were necessary to find good solutions within a reasonable execution time. After much experimentation with various back-jumping and forward checking algorithms, it was found that the quickest results were obtained with standard backtracking where backtracking is somewhat restricted.

It was found empirically that the constraints should be evaluated in the following order:
- the trial time block should be in the specified time zone
- the classroom should be of the proper room type
- the time-classroom pair should not violate any constraints and the constraints should be evaluated in the order shown above.

If no constraints are violated, the course is bound to that classroom during that time block. If not, backtracking is performed with different classrooms and different time blocks. It may happen that no solution can be found even after attempts using all suitable time blocks and classrooms. The secondary constraints have been arranged in a constraint hierarchy[15] and if a course cannot be timetabled, these constraints are relaxed one by one and a new attempt is made. The highest numbered constraint, number 19, is the first constraint to be relaxed *i.e.* has the lowest priority. The lowest numbered of the secondary constraints has the highest priority and is the last to be relaxed.

There is a possibility that even after all secondary constraints are relaxed, a course is still not scheduled. If this occurs, the program enters a procedure that we call *equivalent reversing*, or ER. When a course becomes unschedulable, the ER first chooses an equivalent course which has already been successfully scheduled. It then deletes the latter's bindings and tries to reschedule it after first scheduling the former. If the former cannot be scheduled after deleting the latter's assignments, the ER will replace the latter's assignments and choose another equivalent course and repeat the procedure. Equivalent courses are those which:
- have enrollments in the same size range
- are in the same time zone
- require the same room type and
- request the same building

To avoid long search paths, nested equivalent reversing is forbidden *i.e.* there is no invoking an equivalent reversal from within an equivalent reversal. In addition, the number of equivalent reversals that can be selected for any course has been limited to 4. Of course, assignments to completely prescheduled courses cannot be deleted. If the program cannot schedule a certain course even after the equivalent reversing procedure, it is placed in a list of unschedulable courses to be manually placed later. The goal is to minimize the number of elements in this list.

A number of experiments have been done using the constraint logic approach by other groups using different sets of constraints, formulations of those constraints, implementation languages and platforms. Four different logic based search strategies have been reported by Guéret et al.[16] which yielded solutions after runs ranging from 2 sec. to more then 48 hours. Their data set consisted of 91 lectures, 42 teachers, 8 rooms and 240 starting times. They used the CHIP (Constraint Handling in Prolog) language.

A version of CHIP was also used by Lajos[17] to timetable more than 1000 courses involving around 2500 sessions for first and second year students at the University of Leeds.

The language Oz was used by Henz and Würtz[18] with a Tcl/Tk derived user interface in experiments at DFKI. They scheduled 91 courses and 34 teachers into 7 rooms in less than a minute on a Sun Sparc 20 running at 60 MHz. An improved timetable was found after an additional 10 minutes.

4. Constraint Logic and Tabu Search

The work described in this section was undertaken to investigate the possible relationships between the constraints of the logic based algorithm and the penalties which constitute the objective function of the tabu search. We also wanted a quantitative measure of the time required to produce a solution when the same data was presented to a program which uses constraint logic and to one which uses tabu search. A second investigation was begun to study whether there was any practical advantage to be gained when the output from a constraint logic program was used as the initial solution to the tabu search.

One would suspect that for an institution having many resources there would be no advantage, apart from the time required to reach a solution, of one strategy over the other. The actual solutions might well be different but they would be equivalent in the sense that they would each satisfy the requirements. When resources become scarce, the situation may be different. The constraint based implementation used is one which relaxes one or more of the secondary constraints, if required, in order to obtain a solution. This implies that when the same problem is formulated as a tabu search, the relaxed constraints will show up as a feature which generates a penalty in the tabu objective function. This penalty will be minimized by the perturbation that the tabu search makes to the trial solutions. Thus by feeding the output of a solution produced by constraint logic to a tabu search program, one will obtain a solution which is no worse than the original one, and may well be better. The solution eventually produced by the tabu search in this way should be produced faster than than would be the case if the tabu search were started with a random initial solution.

The data used for this study consisted of several small data sets based on the offerings of certain departments of the Faculty of Science in the Fall semester of 1996. The week was partitioned into 5 days, each one starting at 8:00 and ending at 22:30. The length of each timeslot, the basic quantum of time in this exercise, was fixed at 1.5 hours. There were 18 constraints used throughout.

Four data sets of various sizes were used. A summary of their important features is shown in Table 1.

Number of courses	140	244	262
Number of rooms	21	46	40
μCTS time (sec)	3.6	44.9	38.5
Penalty before tabu	42	50	46
Penalty after tabu	42	42	0
Tabu time (sec)	380	3948	9501
Tabu time/μCTS time	106	88	247

Table 1

The most striking feature of the data in table 1 is the ratio of the times required to terminate the two algorithms. The tabu search algorithm requires approximately two orders of magnitude more time to terminate then does the constraint logic algorithm. Recall that that these measurements were performed by using the constraint logic program's output as tabu's initial solution.

The constraint logic program often produces solutions that have relaxed one or more of the constraints. The more courses there are to be scheduled and the fewer the resources available to schedule them in, the more likely it is that one or more constraints will be relaxed. The tabu search algorithm can reduce the total penalty generated by this relaxation if enough time is available for it to terminate. In one of the cases investigated, the total penalty of a timetable was reduced from 50 to 42, and in another from 46 to 0. In the latter case, it required 247 times as long for the reduction to 0 as it did to produce the timetable in the first place.

5. Tabu Search Alone

One run was made using the tabu search program alone without any aid from μCTS. The initial solution was taken to be essentially random; the rest of the program was unchanged. A summary of this run is shown in the first numerical column of table 2. The objective function had an initial value of 1126 and converged to 0 in about 1 day.

For comparison the same data with the same requirements were used to cast a solution using constraint logic. A solution was found by the constraint logic search alone in 4.8 seconds. A summary of this run is shown in the second numerical column of table 2. The output of this run was used as input to the tabu program as was the case for the data of Table 1.

	tabu alone	μCTS & tabu
Number of courses	104	104
Number of rooms	169	169
μCTS time (sec)		4.8
Penalty before tabu	1126	368
Penalty after tabu	0	0
Tabu time (sec)	≈86400	3780
Tabu time/μCTS time		788

Table 2

It is of interest to note that an optimal solution was finally obtained although the time taken to get it was long. In the end, there were no violations of any of the constraints. The data set in this example was constructed deliberately such that many of the secondary constraints regarding timezone and building preferences could not be satisfied by the logic program. In this case, this complexity was not sufficient to cause any lasting penalty.

6. Comparison

It is evident that starting the tabu search with an initial solution that is known to be good greatly reduces the time required to find the optimum solution, in this case by a factor of about 23. It is also interesting that the final solution in both cases was such that the value of the objective function was equal to 0. The solutions were not identical. This behaviour is obtained for this set of data because the number of resources was quite large and a solution could be obtained without excessive difficulty. This result confirms an observation made earlier by Schaerf and Schaerf[12] .

The tabu search algorithm spends most of its time escaping from regions that are unsatisfactory when starting from a random position. It appears to have difficulty doing so. The rate of decrease in the value of the objective function is small at first but increases with time as better and better solutions are investigated. The difficulty of escaping the really unsatisfactory regions of the solution space is caused by the fact that in the neighbouring regions of the starting solution, improvement is small and the rate of progress is slow. As better regions are found, the rate of escape increases and improvement gets better and better as time passes.

Still, the fastest method by far is the constraint logic method where solutions are routinely obtained in less than one minute, even if they are not as good as those obtained by the tabu search. It is questionable whether further processing by the tabu search algorithm after a solution is obtained by constraint logic is justified in view of the much longer time required.

More experiments are being performed on additional data.

7. Conclusion

The best solution to the timetabling problem for the data sets used was obtained by a sequential use of the constraint logic program followed by the tabu search program. This combination usually produces a timetable with a lower penalty than that of a timetable produced by constraint logic alone. The combination also terminates much faster than tabu search alone. The fastest solution of all was obtained by a constraint logic program used alone.

For the data sets used, the improvements in the timetables achieved by the use of the tabu search program usually do not seem to be worth the extra time involved.

References

1. Victor A. Bardadym, "Computer-Aided School and University Timetabling: The New Wave," *Lecture Notes in Computer Science*, vol. 1153, pp. 22-45, 1996.

2. Michael W. Carter and Gilbert Laporte, "Recent Developments in Practical Examination Timetabling," *Lecture Notes in Computer Science*, vol. 1153, pp. 3-21, 1996.

3. M.W. Carter, "A Survey of Practical Applications of Examination Timetabling Algorithms," *Operations Research*, vol. 34, no. 2, pp. 193-202, Mar-Apr. 1986.

4. Richard A. Dempsey and Henry P. Traverso, *Scheduling the Secondary School*, Nat. Assoc. of Secondary School Principals, Reston, Virginia, U.S.A., 1983.

5. J.E. Brookes, *Timetable Planning*, Heinemann Educational Books, London, U.K., 1980.

6. David Corne and Peter Ross, "Peckish Initialisation Strategies for Evolutionary Timetabling," *Lecture Notes in Computer Science*, vol. 1153, pp. 227-240, 1995.

7. Fred Glover, "Tabu Search: A Tutorial," *Interfaces*, vol. 20, no. 4, pp. 74-94, Jul-Aug. 1990.

8. D. de Werra, "Some Combinatorial Models for Course Scheduling," *Proc. of the 1st Int. Conf. on the Practice and Theory of Automated Timetabling*, (ed). E.K. Burke and P. Ross, pp. 1-20, Napier University, Scotland, 30 Aug. - 1 Sep. 1995.

9. Czarina Cheng, Le Kang, Norrus Leung, and George M. White, "Investigations of a Constraint Logic Programming Approach to University Timetabling," *Lecture Notes in Computer Science*, vol. 1153, pp. 112-129, 1996.

10. A. Hertz, "Tabu Search for large scale timetabling problems," *Eur. J. Op. Res.*, vol. 54, pp. 39-47, 1991.

11. Daniel Costa, "A Tabu Search Algorithm for Computing an Operational Timetable," *Eur. J. Op. Res.*, vol. 76, no. 1, pp. 98-110, Jul. 6, 1994.

12. Andrea Schaerf and Marco Schaerf, "Local Search Techniques for High School Timetabling," *Proc. of the 1st Int. Conf. on the Practice and Theory of Automated Timetabling*, (ed). E.K. Burke and P. Ross, pp. 313-323, Napier University, Scotland, 30 Aug. - 1 Sep. 1995.

13. Pascal van Hentenryck, *Constraint Satisfaction in Logic Programming*, The MIT Press, Cambridge, Massachusetts, 1989.

14. Le Kang and George M. White, "A Logic Approach to the Resolution of Constraints in Timetabling," *Eur. J. Op. Res.*, vol. 61, no. 3, pp. 306-317, 1992.

15. Alan Borning, Bjorn Freeman-Benson, and Molly Wilson, "Constraint Hierarchies," *Lisp and Symbolic Computation*, vol. 5, pp. 223-270, 1992.

16. Christelle Guéret, Narendra Jussien, Patrice Boizumault, and Christian Prins, "Building University Timetables Using Constraint Logic Programming," *Lecture Notes in Computer Science*, vol. 1153, pp. 130-145, 1996.

17. Gyuri Lajos, "Complete University Modular Timetabling Using Constraint Logic Programming," *Lecture Notes in Computer Science*, vol. 1153, pp. 146-161, 1996.

18. Martin Henz and Jörg Würtz, "Using Oz for College Timetabling," *Lecture Notes in Computer Science*, vol. 1153, pp. 162-177, 1996.

Graph Theory

Construction of Basic Match Schedules for Sports Competitions by Using Graph Theory

Arjen van Weert[1], Jan A.M. Schreuder[2]

[1] Agrotechnological Research Institute ATO-DLO, P.O. box 17,
NL-6700 AA Wageningen, The Netherlands
a.vanweert@ato.dlo.nl
[2] University of Twente, Faculty of Mathematical Sciences, P.O. box 217,
NL-7500 AE Enschede, The Netherlands
j.a.m.schreuder@math.utwente.nl

Abstract. Basic Match Schedules are important for constructing sports timetables. Firstly these schedules guarantee the fairness of the sports competitions and secondly they reduce the complexity of the problem. This paper presents an approach to the problem of finding Basic Match Schedules for sports competitions. The approach is clarified by applying it to the Dutch volleyball competition. By using graph representation and theory this approach has the potential to classify sports competitions and to build libraries with Basic Match Schedules for each class of sports competitions. As an example we present an overview of some Basic Match Schedules for the volleyball competition.

1 Introduction

Constructing sports timetables used to be a task for volunteers. These volunteers used their own approaches and had to face just a few constraints. This situation is changing. Sport leagues are organized more professionally and more commercial interests are involved. For this reason, sport leagues are demanding structural methods to construct their timetables.

Constructing sports timetables is a complex task. Constraints are often closely related to each other. Above all, many potential timetables have to be checked in most cases. By using computers this can be performed more accurately. Some examples of computerized timetabling can be found in [1] or [10].

To handle the complexity it is common practice to decompose the construction of sports timetables. This construction is usually decomposed in two phase's [5]. In the first phase a Basic Match Schedule (BMS) is constructed. A BMS determines, for each competition round, the matches and groups the teams have to play in and thus gives the technical outline of the specified sports competition. In the second phase individual teams are assigned to the matches and groups of the constructed BMS. In this phase constraints connected to the individual teams are taken into account. Each phase needs a specific approach.

In this paper we will focus on the construction of Basic Match Schedules for the Dutch volleyball league. In Section 2 we present the characteristics of the Dutch volleyball competition. Section 3 presents the general approach used to find Basic Match Schedules for the volleyball competition. The execution of this approach is presented in Section 4. Section 5 concludes the paper and indicates how the approach can be used for the classification of all sports competitions.

2 The Dutch Volleyball Competition as Example

A volleyball competition consists of a number of sport teams competing with each other. The volleyball competition is subjected to competition rules and external wishes. The competition rules guarantee the fairness of the competition. The external wishes guarantee more or less the practicability of the competition and are focused on specific teams.

The following rules and wishes apply for the Dutch volleyball competition [11]:

Competition rules
1. Each team plays against two opponents in one group in each round.
2. The group sizes are fixed and are the same in each round.
3. The competition consists of two halves. In each half the teams play once against each other.
4. The second competition half is a copy of the first competition half.

The group structure is based on the fact that in a round a group of teams meets in a specific sport venue to play their matches.

Possible external wishes
1. Two teams do not want to play against each other in a specific round.
2. A specific team wants to play in a specific round close to home.

The number of external wishes tends to increase every year. The wishes expressed above are examples of the vast amount of possible external wishes.

In the following Sections we show how to construct a Basic Match Schedule for the described volleyball competition.

3 Construction of Basic Match Schedules

First we present a definition for a Basic Match Schedule.

Definition 3.1
A *Basic Match Schedule (BMS)* determines for each competition round the matches and groups the teams have to play in. A Basic Match Schedule is denoted as a four

dimensional matrix W, where W_{ijkl} defines the opponent k for team i in round j played in group l.

Table 3.1. Example of BMS for round 1 and group 1

4 teams, 1 round and 1 group (W_{ijkl})		
Team	Opponent 1 ($k=1$)	Opponent 2 ($k=2$)
1	2	3
2	1	4
3	4	1
4	3	2

Starting from the Dutch volleyball competition description, we use a three-step approach to reach a Basic Match Schedule. This approach is depicted in Figure 3.1.

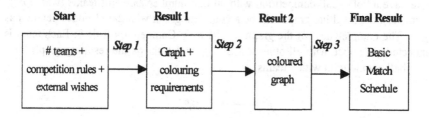

Figure 3.1: Three-step approach to the construction of Basic Match Schedule.

This approach starts with an instance of one half of the volleyball competition as defined in Section 2. The second half is copied from the second half as stated in Competition Rule 4. For the Basic Match Schedules the external wishes are ignored. Only the competition rules are taken into account.

Step 1: Represent the competition in graph
In Step 1 the teams and matches are represented in a graph. Each team is represented in a graph by a vertex. Each match between two teams can be represented by an edge between these vertices. The advantage of this graph is the compact representation of teams and matches. Graph models are already used for other timetabling problems. For an example, see [8] or [9].

Step 2: Represent the rounds in a graph
By colouring the edges of the graph constructed in Step 1 the rounds for the matches can be indicated. A specific round is represented by one colour. We can assign each edge to one of these colours. This colour indicates the round of the match related to the edge. Depending on the characteristics of a sport competition (e.g. group sizes) certain graph theoretic properties will hold for the graph colouring. These properties

can be used to prove that a BMS can be constructed or can help to construct a BMS.

Step 3: coloured graph to BMS
The resulting total colouring of Step 2 represents a BMS.

This approach is clarified by applying it to the volleyball competition.

4   Results for Volleyball

4.1   Odd number of teams

Step 1
If we have a volleyball competition with an odd number $2m+1$ of teams (where m is an integer), the resulting graph will be a complete graph with $2m+1$ nodes denoted as K_{2m+1}. The completeness of the graph results from Competition Rule 3: Each team is connected only once with all other teams. Figure 4.1 shows the resulting graph for a volleyball competition with 7 teams.

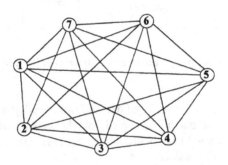

Figure 4.1. complete graph for 7 teams

Step 2
The edges of the complete graph resulting from Step 1 must be coloured with m colours, each colour representing one round. Competition Rules 1 and 2 do not allow just any colouring. A graph induced by one colour must be spanning and 2-regular. This means that all vertices must be connected with two other vertices by two edges of that specific colour. A graph with such properties is an example of a 2-factor. In general a 2-factor of a graph is defined as a 2-regular spanning subgraph. Each group of k teams must be represented in the 2-factor by a set of cycles spanning k vertices. Note that for $k < 6$ this set of cycles is always reduced to one single cycle. The results in this paper are based on single cycles for all k.

We show that only for a competition with an odd number of teams it is possible to find such a colouring.

Theorem 4.1
If K_n can be coloured resulting in edge-disjunct isomorphic 2-factors, then n is odd.

Proof.
Let F_k be a 2-factor for K_n. Then the number of edges in F_k is n. The total number of edges in K_n is $n(n-1)/2$, so $(n-1)/2$ edge-disjunct 2-factors can be coloured. This proves that n must be odd.

In Figure 4.2 the fat lines indicate one round in the graph of Figure 4.1. One group of four teams and one group of three teams are represented.

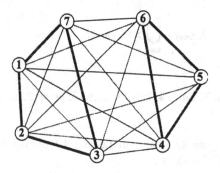

Figure 4.2. Round with group sizes 4 and 3.

We formulate the problem of constructing a BMS for sport competitions with an odd number of teams as the graph theoretic Oberwolfach problem.

Definition 4.1
The *Oberwolfach problem* is defined as finding for a K_{2m+1} a splitting of the edges in m coloured edge-disjunct isomorphic 2-factors F_k. Each F_k consists of vertex disjunct cycles $C_{k1}, C_{k2},.., C_{ks}$ where $k_1 + k_2 + .. + k_s = 2m + 1$, k_1 to k_s are the group sizes. The problem is denoted as $OP(k_1,..k_s)$.

The number of rounds is equal to m. The number of teams in group i is equal to k_i $(i=1,..,s)$. For some m, a construction of these 2-factors is known [2]; for some m it is known that no construction exists. As an example we will show an unpublished construction of 2-factors for $OP(4,4,5)$ in a structural way.

We use lemmas from [4]. These lemmas show how permutations of vertices can be used to find a union of disjunctive cycles for classes of complete graphs.

Notations:

i: Integer

σ_i : Permutation *i* of vertices

$\sigma_i = (1,2,3,4)(5)$ means vertex 1 is permuted to vertex 2, vertex 2 is permuted to vertex 3, vertex 3 is permuted to vertex 4, vertex 4 is permuted to vertex 1 and vertex 5 is permuted to vertex 5.

C_i: Cycle of *i* vertices

K_i: Complete graph on *i* vertices

$K_{i,j}$: Complete bipartite graph with two independent sets of *i* vertices and *j* vertices: All possible edges between the sets of vertices are present, within the sets of vertices no edges are present.

Lemma 4.1. For all *m* :

$$K_{2m+1} = \bigcup_{h=0}^{m-1} \sigma_1^h (C_{2m+1})$$

with $C_{2m+1} = (1;2; 2m; 3; 2m\text{-}1; 4; 2m\text{-}1;5; \dots ;m+3; m; m+2; m+1; 2m+1)$
$\sigma_1 = (1,2,..,2m)(2m+1)$

In Figure 4.3 and Figure 4.4 results of Lemma 4.1 are depicted for $m = 2$, $C_5 = (1;2;4;3;5)$ and $\sigma_1 = (1,2,3,4)(5)$.

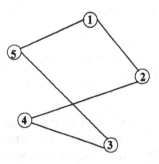

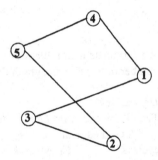

Figure 4.3. $\sigma_1^0 (C_5) = C_5$ **Figure 4.4.** $\sigma_1^1 (C_5)$

Lemma 4.2. For all m,n :

$$K_{nm,nm} = \bigcup_{j=0}^{n-1} \sigma_2^j \; (\; K_{m,m}^1 \cup ... \cup K_{m,m}^n)$$

With $\sigma_2 = (1,m+1,2m+1,..,(n-1)m+1)(2,m+2,..,(n-1)m+1)..(m,2m,..,nm)(nm+1)..(2nm)$

and $K_{m,m}^i := \{\{a,b\} \mid a \in A^i, b \in B^i\}$
with $A^i := \{(i-1)m+1,..,im\}$, $B^i := \{nm+x \mid x, A^i\}$

Lemma 4.3. For all m,n :

$$K_{m(2n-1)+1} = \bigcup_{k=0}^{2n-2} \sigma_3^k (K_{m+1} \cup K_{m,m}^1 \cup .. \cup K_{m,m}^{n-1})$$

With $K_{m+1} := \{1,2n,2(2n-1)+1,3(2n-1)+1,..,(m-1)(2n-1)+1, m(2n-1)+1)\}$,
For $K_{m,m}^i$, $\{e_1^i, e_2^i,..,e_{2m}^i\}$ is the set of vertices,
where $e_{2k-1}^i := (k-1)(2n-1)+i+1$,
$e_{2k}^i := k(2n-1)-i+1$,
and $\{\{e_\nu^i, e_\mu^i\} \mid \nu+\mu \equiv 1 (\text{mod } 2)\}$ is the set of edges
$\sigma_3 = (1,2,..,2n-1)(2n,..,2(2n-1))...((m-1)(2n-1) + 1,..,m(2n-1))(m(2n-1)+1)$.

With the lemma's it can be verified that

$$K_{13} = \bigcup_{j=0} \sigma_3^j ((\bigcup_{i=0} \sigma_1^i (C_5)) \cup (\bigcup_{k=0} \sigma_2^k (K_{2,2} \cup K_{2,2})))$$

with $K_{2,2}$ isomorphic to C_4

From this it follows that K_{13} is the union of the following cycles:
(1;4;7;10;13) (3;5;9;11) (6;2;12;8) $(= F_1)$
(4;10;1;7;13) (6;5;12;11) (3;2;9;8) $(= F_2)$
(2;5;8;11;13) (1;6;7;12) (4;3;10;9) $(= F_3)$
(5;11;2;8;13) (4;6;10;12) (1;3;7;9) $(= F_4)$
(3;6;9;12;13) (2;4;8;10) (5;1;11;7) $(= F_5)$
(6;12;3;9;13) (5;4;11;10) (2;1;8;7) $(= F_6)$
Each row represents one 2-factor, so K_{13} is the union of six 2-factors.
In Table 4.1 all known constructions are listed for K_n ($n < 16$), with n odd.

Table 4.1. Overview of constructions for the Oberwolfach problem

Number of teams	Group sizes	Construction possible?	Reference
7	3,4	yes	[4]
9	3,3,3	yes	[3]/[4]
	4,5	no	[4]
11	3,3,5	not known	
	5,6	yes	[4]
	3,4,4	yes	[4]
13	6,7	yes	[4]
	5,8	yes	[4]
	4,9	yes	[4]
	3,10	not known	
	4,4,5	yes	This paper
	3,5,5	not known	
	3,4,6	not known	
	3,3,7	yes	[4]
	3,3,3,4	not known	
15	all combina-tions	yes	[4]

Step 3

As stated in Step 2 each coloured edge in the graph represents a match in a round and the extra graph properties guarantee the group sizes. This means that each construction for the Oberwolfach problem can be translated to a BMS for a specific volleyball competition.

As an example, Table 4.2 and Table 4.3 present two groups in the first round for OP(4,4,5), resulting from the 2-factors presented in Step 2.

Table 4.2. W_{i1k1}, BMS for round 1 and group 1

Team	Opponent 1	Opponent 2
1	4	13
4	1	7
13	10	1
7	10	4
10	7	13

Table 4.3. W_{i1k2}, BMS for round 1 and group 2

Team	Opponent 1	Opponent 2
2	6	12
6	2	8
12	8	2
8	12	6

4.2 Even number of teams

Step 1
If we have a volleyball competition with an even number of teams, the resulting graph will be a complete graph with $2m$ nodes denoted as K_{2m}.

Step 2
Following from the results of Step 2, Paragraph 4.2 there is no colouring possible obeying all competition rules. To find a BMS for this specific competition with an even number of teams we have to define slightly different competition rules. The rounds are divided into *normal* rounds and one *intermediate* round. In a normal round each team plays two matches against different opponents within a group. In the intermediate round each team plays only one match. After all normal rounds and the intermediate round each pair of teams has encountered each other once.

We will formulate the problem of constructing such a BMS as the problem to find for a K_{2m} a splitting of the edges in $(2m-2)/2$ coloured isomorphic edge-disjunct 2-factors F_k and a perfect matching. Each F_k consists of vertex disjoint cycles $C_{k1}, C_{k2},..., C_{ks}$ where $k_1 + k_2 + .. + k_s = 2m$. The problem is denoted as $KP(k_1,k_2,..,k_s)$. $2m$ is equal to the number of teams, k_i $(i=1,...s)$ are equal to the number of teams in group i. The perfect matching corresponds to the intermediate round.

It can be proved for a K_{2m} that if $(n-2)/2$ isomorphic edge-disjunct 2-factors exist, a perfect matching is left over. In Table 4.4 all known constructions are listed for K_n ($n < 16$), with n even.

Step 3
Step 3 can be performed in the same way as Step 3 for an odd number of teams.

Table 4.4. Overview of constructions for the KP problem

Number of teams	Group sizes	Construction possible?	Reference
6	3,3	no	[3]
8	4,4	yes	[7]
	3,5	unknown	
10	5,5	unknown	
	3,4,3	yes	[7]
12	6,6	unknown	
	5,7	unknown	
	4,8	unknown	
	3,9	unknown	
	4,4,4	yes	[7]
	3,4,5	unknown	
	3,3,6	unknown	
	3,3,3,3	no	[3]
14	all combinations	unknown	

5 Discussion and Conclusion

The presented three-step approach makes it possible to find Basic Match Schedules for the defined class of volleyball competition problems. These Basic Match Schedules are used successfully by the organizers of the Dutch volleyball competition. In our opinion, this approach is applicable to many sports timetabling problems.

In many sports competitions home and away patterns for the matches must be met and multiple matches between two teams have to be scheduled. In these cases, a directed graph model or a multi graph model can be explored as in Step 1. In Step 2 the graph properties have to be defined. In our case of the volleyball competition, the 2-factor properties are used. Comparable results are expected for football competitions [5]. In this way, sports competitions can be divided into classes each with its own graph properties and results. Using these classes in libraries will allow quick construction of Basic Match Schedules for new competitions. This will give timetable constructors more time to invest in the second phase of constructing sports timetables: Assigning individual teams to the Basic Match Schedules subjected to external wishes [6].

Future research will focus on the unknown constructions and on the analysis of other sports competitions.

References

[1] R.R. Bakker et al., Diagnosing and solving over-determined constraint satisfaction problems, Chambéry, 1993, proceedings IJCAI-93.

[2] S. Brandt, Packen und einbetten von Graphen, Freie Universität Berlin, MSc report (in German), 1991.

[3] R. Häggkvist, Decompositions of complete bipartite graphs, Surveys in combinatorics, 1989, pp. 115-147.

[4] E. Köhler, Über das Oberwolfacher Problem, 1977, pp. 189-201, proceedings des symposiums über geometrische algebra (in German).

[5] J.A.M. Schreuder, Construction of fixture lists for professional football leagues, University of Strathclyde, PhD thesis, 1993.

[6] J.A.M. Schreuder, Constraint Satisfaction for Requirement Handling of Football Fixture Lists, Belgian Journal of Operations Research, vol. 35, no.2, pp. 51-60, 1995.

[7] A. van Weert, A solution strategy for constructing sport timetables, University of Twente, MSc report (in Dutch), 1994.

[8] D. de Werra, Some combinatorial Models for Course Scheduling, 1996, pp. 296-308, Lecture Notes in Computer Science (1153).

[9] D. de Werra, Some Models of Graphs for Scheduling Sports Competitions, Discrete Applied Mathematics, 21, pp. 47-65, 1988.

[10] R. Willis, Scheduling the Australian State cricket season using simulated annealing, J.Opl.Res.Soc., vol. 45, no. 3, pp. 276-280, 1994.

[11] A.C.S. Wisse, Automatisering van een sportorganisatie, University of Twente, MSc report (in Dutch), 1992.

Practical Issues

A Standard Data Format for Timetabling Instances

Edmund K. Burke[1], Jeffrey H. Kingston[2], Paul A. Pepper[1]

[1] Department of Computer Science,
University of Nottingham, UK
{ekb, pap}@cs.nott.ac.uk
[2] Basser Department of Computer Science,
University of Sydney, Australia
jeff@cs.usyd.edu.au

Abstract. Many formats have been developed for representing timetable data instances. The variety of data formats currently in use makes the comparison of research results and exchange of data extremely difficult. As many researchers have observed, the need for a standard data format for timetabling instances is now becoming urgent, so as to allow the sharing of data and testing of algorithms on standard benchmarks. At present this is not possible in research on timetable construction. In this paper we state the general requirements of a standard data format and present a standard data format and language for use in the evaluation of timetable instances.

1 Introduction

Research into computerised timetabling has been carried out since the early years of the computer age. A review of the vast literature on this subject [1, 8, 9, 10, 11] quickly reveals that most of the work was undertaken in ignorance of the work of others, and cases where one researcher has produced results directly comparable with the results of another are extremely rare. Consequently, it is difficult to evaluate the contributions made by practically oriented researchers.

Recently, the advent of a series of conferences devoted to computerised timetabling [2, 3] has brought the timetabling research community together and stimulated much fertile exchange of ideas. There has also been some exchange of university examination timetabling data (e.g.[4]) between different institutions in different countries. But the problem of evaluating practical contributions remains, and will do so until the timetabling systems of different researchers are routinely tested on common data.

Other research communities have solved this problem of sharing data. For example, research into the travelling salesperson problem has been facilitated for many years now by TSPLIB [5], a library of instances of the travelling salesperson problem which is freely available over the internet and universally used as test data for programs which attempt to solve this problem.

A prerequisite for any such library is a standard data format with an agreed meaning. Of course, instances of the travelling salesperson problem are very simply described, as graphs whose edges have numeric weights. Real-world instances of the timetabling problem are much more complex. There are many different constraints, their precise nature varies subtly, and they are seldom stated explicitly in data files. Describing these constraints is a difficult problem.

The only previous proposal for a standard data format for timetabling known to us is a discussion paper presented by Andrew Cumming at the 1st International Conference on the Practice and Theory of Automated Timetabling (held in Edinburgh in August/September 1995), but not submitted formally to the conference or printed in its proceedings. That proposal was criticised for lack of generality, but it was valuable in stimulating a discussion during which the full difficulty of the problem became apparent.

We should stress here that the format we propose is intended for storing data centrally, facilitating the exchange of timetable data between researchers via this central format, and allowing more objective means of comparing and assessing timetable solutions. It is not our intention to provide a language from which programs are written to create timetable solutions. A researcher's proprietary format might be more suitable when creating timetable solutions to problem instances.

In this paper we propose a standard data format for instances of the timetabling problem. Section 2 presents the requirements that such a format must satisfy, and Section 3 presents the proposed format itself. Like all standards, it will be successful only if it meets the needs of its users, and accordingly the authors welcome all comments and suggestions.

2 Requirements

This section presents the requirements that have guided our design of a standard data format for timetabling.

The first requirement is generality. It must be possible to express any reasonable instance in the format, for otherwise it will be unacceptable to those whose instances are excluded. It is plainly impossible to enumerate every constraint that researchers have encountered or will encounter. Instead, we propose a format based on set theory and logic, in which any computable function may be defined.

The second requirement is that it must be possible to express instances completely. This includes straightforward data expressing the resources and meetings, complex logical constraints, and proposed solutions. It also includes the criteria by which solutions are to be judged: The hard and soft constraints, and the weighting of the soft constraints, must be explicitly given in the instance itself. Otherwise there will be no agreement on the relative quality of proposed solutions.

This requirement offers an opportunity that we intend to exploit. It will be possible to evaluate any proposed solution against the criteria laid down in the instance. We propose to make freely available a computer program to do this, and thus to establish an independent means of evaluating and comparing proposed solutions.

The third requirement is that translation each way between the standard format and the many existing formats used by researchers must be practicable.

The complex set theory and logic expressions required to express constraints precisely cause a double difficulty here. Not only are they themselves non-trivial to parse and translate, they will in general have no analogues in the data formats of individual researchers, since, as remarked earlier, such constraints are rarely made explicit.

Both difficulties may be overcome in practice by a format which permits all the set theory and logic expressions to be confined to library files. An instance file may then begin with an include directive referring to such a library file, and serving as the definition of all the complex constraints, and then proceed to give the more straightforward data describing the resources and meetings of the instance. A timetabling system whose built-in assumptions conform to the definitions in the library file need never read that file.

The types of problem tackled by researchers are often so different that interchange of data is inherently impossible. These distinct types of problem will need distinct, incompatible library files. But we are hopeful that data exchange will be possible between researchers whose types of problem, while not exactly identical, are sufficiently similar; and even without data exchange, the advantages of precise definition and independent evaluation of solutions will remain.

3 Proposed Standard Format

In this section we give an outline of the standard data format, including a description of the data types, keywords and the grammatical form of the language. We also outline some further facilities which we intend developing, for use with the language, to simplify the task of the data conversion and testing. Amongst our discussion of the language we demonstrate the composition of some timetabling constraints using our language.

Researchers familiar with the Z specification language [6] will notice that some of the concepts and constructs which we use are similar to those found in Z. This is especially true of the way in which Z uses sets since we felt that it meets our requirements very well. We would therefore like to acknowledge the designers of the Z specification language.

3.1 Outline of the Format

The language which we propose is functional in nature, allowing us to express the timetable problem, in its mathematical form, as simply as possible. Data types catered for in the format include the following; class, function, set, sequence, integer, float, boolean, char, and string.

For the data types integer, float, boolean, char and string the usual operations are defined (e.g. addition, subtraction, multiplication and division for the integer data type).

Classes are provided to allow abstract data types to be created - this will be central to the creation of the component data structures of the format. Variables are placed within the body of a class along with functions (which are likely to be forms of constraint) and should be provided to operate on those variables. The following is a simple example which we shall use to illustrate the class construct within the language.

```
class Meeting
{
    Students  : set of student;
    Room      : room;
    TimeSlot  : time;

    function SufficientSeating() : boolean;
};
```

Here we see that the data types student, room and time, which would also have been defined using the class construct, are aggregated within a new abstract data type Meeting. We could now go on to construct an instance of Meeting in the following manner:

```
Maths1 : Meeting;
Maths1.Students = {Greaves, Watt, Dickens};
```

Notice that the . (dot) operator is used to access the components within the Maths1 object. In this case no values are assigned to the Room and TimeSlot components since these would be used to hold the values which determine a solution to the timetable problem instance (this would be carried out by some form of timetabling application).

Alternatively we could have assigned values to the component objects within the Maths1 object in a single expression using square brackets to contain the assignments:

```
Maths1[ Students = {Greaves, Watt, Dickens},
        Room = ?,
        TimeSlot = ? ];
```

The ? (question mark) operator is used to indicate that no value has yet been assigned to the Room and TimeSlot variables.

Although the example we have chosen to use is a simplification of how a real meeting might be modelled it would be possible to extend our example definition using single inheritance. This would involve using the inherit keyword in the following manner:

```
class ExamMeeting inherit Meeting
{
    ExamId : integer;
}
```

The inheritance mechanism will provide an important feature in allowing standard definitions within the format to be extended where necessary. All standard definitions using the class construct will be placed within library files - a standard meeting class would be an obvious candidate for inclusion within the libraries. These will then be included by a timetabling instance where the inherit keyword may be used to extend any of the classes contained therein.

Our new abstract data type may also require new member functions and the prototypes for these functions could also have been added to our definition of the ExamMeeting using the notation shown in our Meeting for the function SufficientSeating(). The prototype in this case tells us that no parameters are passed in to the function and that its evaluation will result in a boolean. The implementation of SufficientSeating would use the following syntax:

```
function SufficientSeating() : boolean
{
    Room.Capacity >= Count(Students);
}
```

The main body of the function is placed between braces, in the case of SufficientSeating() above the returned value will be true if the capacity of Room is greater than or equal to the number of elements within the set Students. If parameters are to be passed into a function then these should appear between the parentheses placed after the function name. If an integer parameter is passed in to the function then it would take the following form:

```
function SufficientSeating( Unavailable : integer ) : boolean
```

Where multiple function parameters are used, they should be separated by commas and if a reference to a member function's calling object instance is required then the self keyword should be used.

Sets will be another very useful data type within the language since many of the components that timetabling instances deal with involve groups of objects (groups of students representing, say, classes of students, groups of classes, etc.) and viewing these groups as sets allows us to express many of the constraints encountered within timetabing problems clearly and simply.

Sets within the language are defined in their conventional mathematical form whereby all the elements of any given set are unique and the usual operations on sets of objects are defined - we also treat internal data types as sets, the member elements of which would be their literal values. We do, however, restrict any object of type set to contain objects of only one data type. Consequently, mixing different types of resource within a set, for example rooms and invigilators, is not permissible.

In addition to the commonly encountered set operations - member, union, intersect, subset, - (set exclusion) and count - we also include the operators forall, exists, sum and prod. The forall and exists operators have the same meaning as their abstract mathematical operators $\forall$ and $\exists$, respectively, whilst the sum and prod operators are used to evaluate the sum and product, respectively, of numeric values within a set. An example implementation of the *No-Clashes* constraint shows the syntax of the forall operator (the syntax of the exists operator is defined in exactly the same format) and illustrates its expressive power.

```
function NoClashes( Meetings : set of Meeting ) : boolean
{
    forall m1 in Meetings
    (
        forall m2 in Meetings | m1 != m2
        (
            ( m1.Time intersect m2.Time == {} ) or
            ( m1.Students intersect m2.Students == {} )
        );
    );
}
```

In this example the variable m1 is bound within the scope of the first instance of the forall operator (i.e. between the operators associated pair of parentheses) and similarly for the variable m2 though this is also bound within the expression following the vertical bar. The vertical bar is optionally used to restrict the set of elements over which forall will operate and in the above example we have used it to restrict comparisons between non-identical meetings. If we now look at the abstract mathematical form of the NoClashes function we can see that the transformation to an implementation can be achieved relatively easily.

$$\text{NoClashes}(Meetings) =$$
$$\forall m_1 \in Meetings \text{ and } time_1, resource_1 \in m_1,$$
$$\forall m_2 \in Meetings \text{ and } time_2, resource_2 \in m_2,$$
$$m_1 \neq m_2 \Rightarrow (time_1 \cap time_2 = \varnothing) \vee (students_1 \cap students_2 = \varnothing)$$

The syntax of the sum and prod operators is defined in a similar manner to forall and exists. As an example usage of the sum operator we may obtain the sum value of all soft constraints imposed by a timetabling instance as follows:

```
sum m in Meetings (m.Soft())
```

Here Meetings is the set of all meetings and Soft() is a member function defined within the class definition of a meeting. The type system would guarantee that Soft() is defined for all instances of the meeting class and this could then be used to evaluate any associated soft constraints, as shown in this example.

Difficulties can arise when defining equality and equivalence (identity) within a language. The implementation of the NoClashes function above provides an example for us to illustrate the rules which we define to distinguish between the equivalence and the equality of variables.

When two variables of a user defined type (the type being created using the class feature) are compared using either of the operators == or !=, we define the operation to be a comparison of equivalence. Although reassignment is not permitted we use the == operator to avoid any confusion with the assignment operator, =. In the NoClashes example above a comparison of the identities of the variables m1 and m2 is carried out. If a programmer wishes to compare the contents of two user defined objects, of the same type, then he should define a comparison function within the body of the class definition to carry this out for him. It is only when an object instance of a user defined type is compared with a literal value that the contents of that object are compared. This kind of comparison can be seen in the NoClashes implementation. An object of type set of Time is compared with the empty set literal value, {}, using the == operator. The comparison of object instances of built in types (such as integer, float, char, etc.) using the operators == and != will involve the comparison of the contents of those instances (i.e. checking the equality/inequality of the objects).

The seq data type is used to model sequences within the language. It is similar to the set data type in that most of the valid operations for set objects have their analogue for seq objects. The usual exceptions to this would be the head and tail sequence operators which are included in the language. The literal values of the seq data type may be expressed in the same format as set literal values. Chevrons are used to delimit the seq literal value where braces would be used to delimit the literal values of the set data type.

In order to express conditionals we use the if construct and illustrate its syntax as follows:

```
if count Maths1.Students > Rooms.MainHall then
    // Return expression if condition holds
else
    // Return some other expression otherwise
end
```

We also include the let keyword in the language to allow commonly used expressions to be abbreviated by a user defined synonym within an expression. The syntax of this will be defined as shown in the following simple example:

```
let Penalty = (Meeting.Room.SeatingCapacity -
               Count(Meeting.Students)) * 0.1
in
    // Penalty is now used within this block as a variable
end
```

As can be seen in many of the examples provided within this section the double forwards slash, //, is used to indicate the start position of a comment on a single line.

3.2 Other Facilities

In addition to the language and standard data format we will also develop a library of standard functions. These will consist of the most commonly encountered constraints found in timetabling and general utility functions likely to be used within timetabling. [7] provides a good starting point for the selection of such constraints. All the standard functions will be implemented using the language and will comply with the standard format. The simplified No-Clashes constraint, which was used as an example in the previous section, would be one such constraint to have its implementation appear in the library. Examples of other constraints which might also appear in the library would include Sufficient-Seating (a hard constraint which states that there should be sufficient seats available within a given set of rooms for a meeting or a selected number of meetings), Share-Period (a constraint stating that a given set of meetings must share the same time period), and Restrict-To-Periods (constraining a meeting, or set of meetings, to a given set of time periods).

Programs will be necessary to convert data in a non-standard format into the standard data format. In order to encourage the development of consistent conversion programs we will publish a standard interface. This might then be used to interface with a common conversion and application which would be used as the standard front-end to the conversion utility. Although there is no necessity to make conversion programs publicly available we would encourage the contribution of data expressed in the standard format and would hope to provide assistance with the conversion process where required.

The conversion of data into the standard data format will allow us to build up a pool of timetabling data, incorporating the variety of constraints encountered within this area. A large range of data may (eventually) be selected from this pool and converted into a researcher's own format for use in testing the effectiveness and robustness of their methods of timetable solution.

Use of a standard data format will also facilitate objective testing and comparison of results between researchers using different data formats. Data would be taken from the data pool and converted into a researcher's own format for use in validating their timetabling methods. Subsequent conversion back to the standard data format would allow objective comparison of the results of the researcher's methods against other solutions. We aim to develop benchmarking and evaluation applications to assist and promote the consistency of such testing.

4 Conclusion

We hope that our work will promote an open exchange of timetabling data between researchers and the objective comparison of research results (the exchange process will be simplified once conversion programs are written). Just as the success of the Travelling Salesman Problem Library has shown, these objectives are achievable.

We will be looking to place all the facilities outlined in this paper on the internet once they are available. Allowing open access to our work, via the development of a timetabling library internet site, will be important if we are to promote the exchange of ideas with respect to the solution of timetabling problems.

Many formats used by timetabling applications will have been designed with concerns for minimising data storage requirements or to facilitate the fast processing of data. This means that such formats would not be suitable as a standard format since generality or readability of data (by humans) would be forfeited. Our aim, however, is to provide generality; allowing the maximum number of data instances to be represented as is practicable. Since we do not expect our format to be used internally by timetabling applications this aim can be met.

An initial investment in time will be necessary from potential beneficiaries of the standard data format. This and the initial delay in the development of inertia in the growth of the timetable data pool may cause a hesitation in some to become involved, though we think the benefits to be gained will encourage researchers sufficiently.

Suggestions regarding the development and improvement of the standard data format and associated facilities are welcomed by the authors. Moreover, a great deal of the success of the standard data format is reliant upon feedback from those who will make use of it and gain the resulting benefits (e-mail may be addressed to any of the authors with the text "TTLIB comments" placed in the *subject* header section).

References

[1]   Jeffrey H. Kingston, *A Bibliography of Timetabling Papers*, URL "ftp://ftp.cs.su.oz.au/jeff/timetabling/".

[2]   E.K. Burke and P. Ross (eds.), *The Practice and Theory of Automated Timetabling: Selected papers from the First International Conference on The Practice and Theory of Automated Timetabling*, Lecture Notes in Computer Science, Vol. 1153, Springer 1996.

[3]   E.K. Burke and M.W. Carter (eds.), *The Practice and Theory of Automated Timetabling II: Selected papers from the 2nd International Conference on The Practice and Theory of Automated Timetabling*, Lecture Notes in Computer Science, Vol. 1408, Springer 1998.

[4]   E.K. Burke, J.P. Newall. R.F. Weare, *A Memetic Algorithm for University Exam Timetabling*, in [2], pp. 241-250.

[5]   Bob Bixby and Gerd Reinelt, *TSLIB - A Library of Travelling Salesman and Related Problem Instances*, September 1992. URL "http://nhse.cs.rice.edu/softlib/catalog/tslib.html".

[6]   Ben Potter, et. al., *An Introduction to Formal Specification and Z*, Prentice Hall, 1991

[7]   E.K. Burke, et. al., *Examination Timetabling in British Universities - A Survey*, in [2], pp. 76-90.

[8]   M.W. Carter and G. Laporte, *Recent Developments in Practical Examination Timetabling*, in [2], pp. 3-21.

[9]   V.A. Bardadym, *Computer-Aided Lessons Timetables Construction. A Survey*. USIM (Management Systems and Computers, 8, pp. 119-126, 1991.

[10]   M.W. Carter, *A Survey of Practical Applications of Examination Timetabling Algorithms*, in Operations Research, 34, pp. 193-202, 1986.

[11]   W. Junginger, *Timetabling in Germany - a Survey*, Interfaces, 16, pp. 66-74, 1986.

Academic Scheduling

James F. Blakesley, Keith S. Murray, Frederick H. Wolf and Dagmar Murray

Space Management and Academic Scheduling
Purdue University
West Lafayette, IN 47907-1128
{jfb,dmurray}@purdue.edu

Abstract. As the cost of higher education rises, it is becoming increasingly important for colleges and universities to examine how well they are meeting student needs. In addition to the quality of instruction being offered, this includes the availability of courses required to fulfill student degree objectives. One of the greatest unanticipated costs facing many college students is having to pay an extra semester's or year's tuition because they were unable to take all of the courses required for graduation within the expected (usually four year) time frame. A primary goal of university academic scheduling processes should, therefore, be to maximize the probability that all students will receive the courses they require to meet their degree requirements in a timely manner.

It is essential for university administrators who are responsible for the creation of the schedule of classes and the assignment of students to these classes to understand the effects of their decisions on the operation of the university. They can be far reaching: from the educational opportunities created or denied students, to the budgetary bottom line of making efficient use of resources. The purpose of this article is to relate some of these effects so that the reader can better meet student needs.

1   The Schedule of Classes

Both aspects of the scheduling process—creation of the schedule of classes and assignment of individual students to classes—are important to maximizing the student's chance of receiving the courses she requires for a degree. The process starts with building a schedule of classes that helps expand the choice of courses available to the student.

The schedule of classes represents a coordination of the staff, facility, and time resources necessary to offer instructional courses. The choice of courses available to a student is a function of how all of these resources are managed. A limited number of faculty are qualified to teach any given course. The number and size of rooms available are also a constraint. Careful attention to curricular requirements and the relationship between the times courses are taught is essential to avoiding potential conflicts. Making full use of available time is also critical.

2 The Academic Week and Scheduling Time Patterns

The length of the academic week and variations in time patterns used in constructing a schedule of classes play important roles in the effectiveness of the academic schedule. The academic week is the set of days and times during which instructional activity occurs. It may include or exclude the noon hour, evenings, or weekends. Some campuses have different academic weeks for programs geared to different populations (e.g. daytime hours on weekdays for "traditional" students and evenings and/or weekends for working members of the community). In this case, the principles discussed below can be applied to each academic week. Overlaps will provide both additional course opportunities and additional resource constraints. Longer academic weeks provide greater opportunity to avoid scheduling conflicts.

Time patterns are the configurations of days and hours to be used in setting up the schedule of classes, such as MWF 9:00. If a standard set of patterns is chosen, with compatible starting and ending times, schedules will fit together more easily. If patterns are dissimilar, more conflicts will occur within a given academic week. Patterns of hours for laboratory classes vary greatly from those for lectures or recitations but should still be chosen wisely to fit together with one another in the best possible manner (e.g. all three–hour labs are scheduled from 8–11 a.m., 11 a.m.–2 p.m., or 2–5 p.m.). Curricula that require many laboratory hours of instruction have space and time frameworks that are more complex and require an academic week with more available time patterns than do curricula without laboratory classes.

Most institutions have a five or ten minute break between class periods, but some have adopted fifteen or twenty minute breaks. The effect of longer class breaks is a major reduction in instructional time if class periods are shortened, or a reduction in the number of class periods available in the academic week. Institutions should carefully examine the rationale behind such policies. Isolated lateness to classes should not precipitate a campus–wide policy of longer breaks when relocating a class might solve the problem. Planning for campus buildings should include consideration of travel times between instructional facilities. While course offerings in specialized curricula may be located on the perimeter of campus, scheduling core courses to outlying rooms is more likely to create difficulties.

3 Scheduling Rationale

While it may seem desirable to build a schedule that accommodates faculty and student time preferences to the greatest extent possible, this generally works contrary to the primary goal of meeting student course needs. Since there is a tendency by all to avoid the least popular days or hours, such a policy effectively collapses the schedule into a shorter academic week and increases the probability of conflicts. Effort is required to broaden the times utilized in order to satisfy the diverse course needs of the total student population.

The tendency to queue at prime times, is neither academically sound nor economically appropriate. Figure 1 offers a simplified illustration. If all courses were taught at one time, a student's choice of courses would be limited to one per term. Consequently, his ability to progress toward a degree would be severely hampered. Staff and facility requirements would also be at a maximum since a different instructor and room would be required for each course and section offered. On the other hand, if all courses were taught at different times, students could select any set of courses offered with no conflicts in hours. Their ability to progress would be limited only by their own capabilities. In addition, staff and facility resources could be minimized. Theoretically, since there would be no time conflicts, one extraordinarily talented professor could teach all courses. Similarly, if one room were suitable for all of the courses offered, only one room would be required.

In general, an institution that operates on a twenty-hour week is much more restrictive in its course selectivity and less effective in its use of staff and space than an institution that operates on a forty-hour week. That is, the effectiveness of the master schedule of classes is dependent on the distribution of classes over as broad a weekly time base as institutional policy will permit. The more usable hours in the academic week, the more effective the system. Such a schedule will provide a greater selectivity of courses and more utilization of staff and space resources.

In practice, neither of the extremes in Figure 1 is a viable plan for scheduling. Even the "distributed" extreme is limited by the number of hours in the academic week as compared to the number of courses offered. There must certainly be more than one course per period if there are more than forty one-hour courses in a forty hour academic week. If there are, say, two hundred courses, at best there will be five courses per period. More typically there would be thirteen three-hour time patterns in a forty hour academic week, requiring that fifteen to sixteen courses be offered each pattern of hours and thus increasing the probability of conflicts between courses. Further, as the time patterns become more varied and complex, such as required for a chemistry course with a one-hour lecture, two separate one-hour recitation sessions, and one three-hour laboratory, it becomes quite apparent that even fifteen to sixteen courses per time pattern would be a bare minimum, and more than likely many more courses will need to be offered at a given period, again increasing the probability of conflicts.

The interrelationship of time patterns, length of academic week, and number of course offerings becomes more apparent as the complexity increases. A liberal arts curriculum with very limited numbers of laboratory periods and many regular class periods can operate with a much greater potential for a wider selection of courses within a given academic week than can a curriculum that has relatively more complex time patterns, such as those found in the scientific and technical fields. Curricula with

CONCENTRATED COURSE TIMES

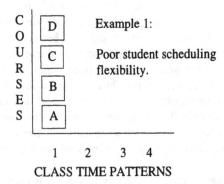

When courses are concentrated at one time, students have minimum course selectivity and no scheduling flexibility. Maximum staff and space resources are required.

DISTRIBUTED COURSE TIMES

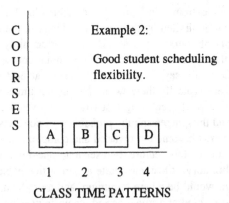

When courses are distributed over the time periods, students have complete course selectivity and scheduling flexibility since any combination of courses may be taken. Staff and space resource needs can be minimized.

Figure 1. A Scheduling Philosophy

complex time patterns require a longer academic week to obtain the same flexibility of course selection as less complicated curricula.

Large single–section courses, especially those required by students from many curricula, must be scheduled with great care. The time selection for such a course will influence schedules in all of the associated curricula and preclude other courses from utilizing the selected time pattern. The direct cost per credit hour for the course may be low, but its affect on other courses may make its true cost to the university prohibitively expensive. One rule of thumb is to encourage at least two time periods for such a course, thus providing scheduling alternatives. In general, single–section courses, especially those with large enrollments, restrict course selection.

Multiple–section courses offer more opportunities for non–conflicting schedules than do the large single–section courses. Scheduling officers should offer as many different standard time patterns as possible for multiple section courses. In most cases, if enrollments in the sections are kept balanced, hour conflicts in student schedules will be minimized. Multiple–section courses, in effect, increase the potential for greater course selectivity.

The planning of the schedule of classes and the ultimate assignment of students to classes are interdependent. A well–conceived master schedule can fail if the more popular sections of multiple–section courses are allowed to fill and close prematurely before all students are registered. As sections close, the flexibility offered by multiple sections is reduced; and as registration proceeds, more and more students find fewer patterns of courses that will fit together to make an acceptable schedule of classes. In practice, therefore, all sections of multiple section courses should be kept open as long as possible by balancing student course requests among all the sections. When this procedure is used, the last student to register will have essentially the same choice of courses as was afforded the first student to register. The penalty for not observing this practice is to deny certain students their choice of courses.

3.1 Theoretical Considerations

To further illustrate the importance of the relationship between the academic week and the time patterns, it is useful to examine the probabilities of developing nonconflicting schedules using a uniform distribution of courses with random selection. Questions are frequently raised about the length of the academic week and the effect that the distribution of hour patterns has on the probability of obtaining a nonconflicting schedule. In answer to the first question, assume a simplified situation in which the academic week is divided into disjoint patterns of uniform length (e.g. MWF 9:00), and where each course meets for exactly one pattern per week. The actual length of a period is immaterial, though an example might be a forty–hour week with eight one–hour periods on Monday, Wednesday, and Friday and five ninety–minute periods on Tuesday and Thursday. In this example; let K equal the number of time patterns available in the academic week, let N equal the total number of courses offered by the institution, and let M equal the average number of courses taken by each student. If it is assumed that the N courses are uniformly distributed over the K time patterns, it is reasonable to ask what the probability would be for a student to randomly select M courses without a conflict. Using the formula for conditional probability

$$P_{\text{no conflict}} = \prod_{i=1}^{M} \frac{N - \frac{N}{K}(i-1)}{N - i + 1}$$

some representative values may be derived, as shown in Table 1.

Table 1. Probabilities of Developing Non–Conflicting Schedules

K = Time Patterns Available	M = Time Patterns to Be Scheduled						
	6	7	8	6	7	8	
10	0.18	0.07	0.02	0.15	0.06	0.02	
11	0.22	0.11	0.04	0.19	0.09	0.03	
12	0.26	0.14	0.06	0.22	0.11	0.05	
13 ----------	0.30	0.17	0.08	0.26	0.14	0.06	◄— 40-hr wk
14	0.33	0.20	0.11	0.29	0.17	0.08	
15	0.37	0.24	0.13	0.32	0.19	0.10	
16	0.40	0.27	0.16	0.35	0.22	0.12	
17	0.43	0.30	0.19	0.37	0.24	0.14	
18 ----------	0.46	0.32	0.21	0.39	0.26	0.16	◄— 55-hr wk
19	0.48	0.35	0.24	0.42	0.29	0.18	
20	0.51	0.38	0.26	0.44	0.31	0.20	
	N = 100 Courses Offered			N = 3,200 Courses Offered			

* Assuming an even distribution of courses with random selection.

A review of these values indicates that the *number of courses offered* has less influence on conflicts than does the *number of time patterns selected* by the students. If there are thirteen time patterns available and the student is choosing six time patterns (courses) from an array of 100 courses, then the probability is 0.30. If the student chooses from an array of 3,200 courses, the probability only drops to 0.26. However, if the *number of courses selected* from a set of 100 courses increases from six to eight, the probability of no conflicts drops significantly, from 0.30 to 0.08. A drop of the same magnitude occurs for the example where the array of courses is 3,200.

These data reaffirm that it is much more difficult to schedule students with "full" academic loads than it is to schedule part–time students within a given academic week. The data also suggest that planning the master schedule is only somewhat more difficult for institutions with many courses than for those with few courses. In each instance, the complexity remains virtually unchanged, but the volume of work increases.

Within a well–constructed schedule of classes, the courses are not randomly arranged, nor do the students randomly select their courses. The probability of obtaining a conflict–free schedule can still be greatly increased when the courses are distributed throughout the hours of the day and the days of the week. In addition, as the number of non–conflicting time patterns increases, the probability of no conflict improves significantly and in an exponential fashion.

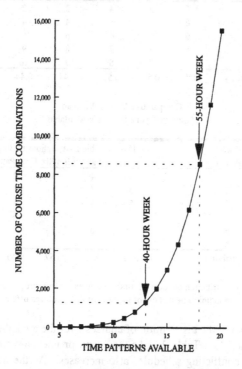

Figure 2. Course–Time Combinations
(Combinations of N courses Taken Five at a Time)

Figure 2 shows the significant increase in possible course–time combinations that occurs as the number of time patterns or hours (N possible courses) in the academic week increases. For example, if there are thirteen discrete periods in the academic week, a student taking five courses would have 1,287 possible course–time combinations. If the academic week is increased to eighteen discrete periods, 8,568 possible combinations would exist for the same student. While calculated for theoretical situations, these data serve to highlight the point that the length of the academic week and the time patterns of hours have a direct and significant impact upon the institution's ability to plan a schedule of classes which meets as nearly as possible the diverse course requirements of students.

**Table 2. Probabilities of Developing
Non–Conflicting Schedules[1]**

Effects of Skewing the Distribution of Courses
Within the Available Time Patterns

Time Pattern	Six Possible Distributions of 30 Course Offerings Within Five Time Patterns					
1	6	5	4	3	2	1
2	6	5	5	4	4	3
3	6	6	6	6	6	6
4	6	7	7	8	9	9
5	6	7	8	9	10	11
Probability	0.53	0.52	0.51	0.47	0.44	0.38

Comparative Values When an
Additional Time Pattern is Available

Time Pattern	Five Possible Distributions of 30 Course Offerings Within Six Time Patterns				
1	5	4	3	2	1
2	5	4	3	3	2
3	5	4	4	4	3
4	5	6	6	6	7
5	5	6	7	7	8
6	5	6	7	8	9
Probability	0.62	0.60	0.57	0.55	0.47

[1]Assuming a skewed distribution of courses. Both examples assume three courses per student will be randomly selected out of the total of thirty courses offered.

The effects of skewing the distribution of courses over the available class periods are also worthy of study (see Table 2.) As the skew (or uneven distribution) increases, the probability of conflicting schedules also increases. As the number of available class periods is decreased, the probability of conflicts increases even further. Therefore, from a course selectivity standpoint, it is highly desirable to distribute courses evenly over an academic week, thus providing for the largest number of nonconflicting time patterns. If choices must be made, courses that are taken in sequence (or otherwise not taken during the same term) may be scheduled at the same time, whereas courses that may be taken during the same term should be scheduled at different times.

3.2   Comments and Suggestions

Lost Weekends: Some institutions are experiencing significant shifts towards shorter academic weeks. This trend, while somewhat expected in institutions with heavy use of Saturday classes, has progressed in some institutions to the point where Friday afternoons (and even Friday mornings) are now void of classes. The results are more scheduling conflicts, longer times to graduation and pressure for additional funds to build classroom facilities. These events may be termed lost weekends. If not observed early and guarded against, this trend is very difficult to reverse since no one wants to be last in line, or on the last period, or on the last half day, etc.

Too Few Better Than Too Many?: Many an administrator has heard the cry for more classroom space. This cry may be quite sincere if the hours of room use are approaching the hours available in the scheduling week, or if there are other special constraints. Care must be taken to sift through these requests to find the reason for the problem. In the general case, however, when there has been a fair allocation of classroom facilities, each department will have an adequate number of time periods for its classes. The availability of additional rooms will not increase the number of classes offered, but it will promote departments being more selective with the time periods they use—resulting in more classes and an increase in conflicts at "prime" times. With fewer classrooms, there will be a better distribution of class times and fewer conflicts.

Centralization of Classrooms: All classrooms should be centrally scheduled and managed, with a firm commitment of adequate resources for their continued upkeep and modernization. This will minimize complaints of differences in quality of rooms about campus. Usually staff will choose to remain and teach in their building at the less popular hours than to move outside the building for a preferred time. Good facilities help to promote this interchangeability of rooms.

Scheduling Large Enrollment Multiple–Section Courses: Take advantage of courses with large enrollments and many sections. These offer a multitude of opportunities and can be scheduled at all time patterns, fill in normal void periods in the master schedule, minimize conflicts in hours and, in general, attain an extremely high level of utilization with minimal conflicts. This is true of both class as well as laboratory sessions. It has also been found that in these high enrollment courses irregular hour patterns such as MTh, TF, and WSa work well and do not increase the level of conflict in hours.

Scheduling Single Section Courses: Take care to use as much of the academic week as possible to minimize possible conflicts. Keep as many courses away from each other as possible within a curriculum to minimize conflicts and only put the most unlikely courses (or known sequence courses) together at the same time.

Scheduling Large Single Section Lecture Courses: These courses can be both economical and expensive. While the credit hour cost may be low, any course required by all students or the majority of them idles all other course activities and facilities on the campus—and should be handled with care and understanding. In the past, ROTC programs had a significant impact on many of the land grant institutions due to the required corps drill period for all male undergraduate students.

Given the same length academic week, an institution with few, if any, multiple-section courses will have a higher level of conflicts between courses than an institution with many multiple-section courses. In other words, given the same level of hour conflicts between courses, multiple-section courses reduce the need for an expanded academic week whereas large single-section courses increase the need for an expanded academic week.

It is important to hold to standard time patterns, but exceptions must be allowed when teaching pedagogy prevails. In this event, request possible alternative time patterns and choose actual time period that best reduces the possibility of conflict.

Scheduling strategies that might be employed:

1. Schedule all multiple lecture and laboratory sections so that student course enrollments will be distributed approximately equally between mornings and afternoons and between the MWF and TTh sequences.

2. Four-hour multiple-section courses with large enrollments meeting four days a week should be scheduled equally over the five following combinations: MTWTh; MTWF; MTThF; MWThF; TWThF.

3. Departments should be strongly encouraged to schedule noon-hour classes, although they should not schedule two required single-section courses or both sections of a two-section course at, say 11:30 and 12:30.

4. Departments adding lecture courses and sections above the total number offered the previous year (term) should schedule the added sections in the low-use time blocks. No increase in peak-period use should be permitted. Likewise, if possible, reductions (cancellations) should come from the peak periods—not the low—volume periods.

4 Conclusion

The creation of an institution's schedule of classes and the policies it uses for assigning students to these classes play a vital role in the ability of its students to receive the courses they need to advance toward their degree objectives. Efforts toward creating an academic schedule which maximizes the students' chance of receiving the courses they need can, therefore, create a cost savings for students as well as more efficient operations for the college or university. This may require some changes in policies and result in a perception of inconvenience by staff and students who have not been scheduled to a preferred time. Students generally appreciate that this is less of an inconvenience than not being able to take a required course however. Effective scheduling policies will also provide some choice of times for students with special needs.

Appendix: Purdue Academic Scheduling System (PASS): Overview

The basic objective of Purdue's student scheduling system is to maximize the probability for all students to receive their first choice of courses which have been selected to meet their educational requirements and interests. Fundamental to the system is the policy of providing students with choice of courses rather than choice of times, and all energies are directed toward scheduling all students, the first as well as the last, into the courses of their first choice. This procedure is in contrast to many registration systems wherein students select both courses and times, usually on a "first come, first served" or on some priority basis.

To accomplish this scheduling objective, it is necessary to first build a Master Schedule of Classes which utilizes all hours of a broadly defined academic week in order to minimize queues of course offerings at any hour. Such a schedule will provide for greater course selectivity than one that queues at popular hours and is restricted to a limited academic week.

Secondly, it is necessary to keep in balance the number of course spaces remaining in the time patterns used by a course throughout the student scheduling process. This procedure prevents the early closing of classes offered at popular times.

Use of standard time patterns campus wide is another important operational policy, which facilitates the development of non-conflicting schedules. When departments teach courses in common blocks of time, fewer time conflicts are likely to arise when building student schedules, thus allowing more students to take the courses they need.

Development of the Master Schedule of Classes

The first step in the scheduling process involves class schedule coordination of academic resources on campus. The coordination of courses, instructional personnel, space, and related resources to develop a workable Master Schedule of Classes within the established operating week and calendar is the joint responsibility of the Office Space Management and Academic Scheduling (SMAS), school counseling offices, and the individual university departments. The objective of this management team is to properly coordinate the schedule of academic resources, instructional personnel, academic space, and time, in a manner, which will effectively meet the curricular needs of the students.

Curriculum deputies represent student counseling areas and are responsible for 1) estimating the demands their students will place on courses in the various departments throughout the university, and 2) coordinating with the departments the establishment of non-conflicting times for courses their students are likely to elect. In essence, the curriculum deputies project the curricular needs of students and act in their behalf to see that the students' needs are met.

Departmental schedule deputies represent subject matter areas and departments of instruction. They are responsible for 1) reviewing the total course demand upon the department in order to balance the requests against available staff and space resources, and 2) developing a compatible departmental schedule within available time constraints. In essence, they manage the department's resources and are responsible for their efficient utilization.

This deputy system of management, is a decentralized form of management, which allows each of the various schools and departments to have a voice in the determination of their unique departmental schedules and the school's establishment of appropriate curricular offerings, while being monitored within the framework of centrally developed scheduling procedures and policies of the university.

The master schedule is reviewed by SMAS, checked for consistency, fairness and adequacy, simulated on the computer to test against actual students requests, modified as needed, and ultimately released as the operating schedule for the University. Adjustments in the master schedule occur throughout the registration period to accommodate student demands and changing departmental requirements.

Scheduling

When about half of the students have registered for a semester, actual scheduling begins. In the scheduling process, students are assigned to the times of their selected courses which have the most open spaces remaining in order to keep the maximum number of time options open for students scheduled later in the registration process.

In order to schedule the most students and to achieve a well-balanced schedule, multiple passes through the student requests are made. To accommodate this, student course requests are collected and periodically submitted for scheduling as an overnight batch procedure.

During schedule revision, all course drops are processed prior to the day's batch of new course requests. This makes class time openings available, giving more students a greater chance of receiving the courses they want.

Although undergraduates are not allowed to select course times except under special circumstances, there are mechanisms that can be used when students have a legitimate need to be free from class for certain blocks of time. If course instructors are known in advance, students may indicate an instructor preference, which will be honored if possible.

In order to protect student's health, the scheduling algorithm attempts to include an hour lunch break in each student's schedule.

Output from the scheduling process includes a schedule for the student who was scheduled or a detailed reject analysis explaining why a student could not be scheduled. Departments are provided with course status reports and summaries of scheduling activity throughout the scheduling process.

Scheduling Vision

Once the schedule of classes is public, students could use computers to enter their course requests and other information. Along with the schedule of classes, additional information about courses, the student's course history and curriculum requirements would be on-line to aid with decision-making.

Once submitted by the students, the request could be forwarded electronically to the student's advisor for approval. The level of interaction with students could be determined by each advisor. Approved scheduling requests would be electronically submitted into the subscription system. Perhaps special permission requirements could be handled in an electronic fashion as well.

Batch scheduling would take place during the advanced registration period. This allows modifications to the master schedule to accommodate as many student scheduling requests as possible, thus allowing more students to take the courses they need.

During the delayed registration period, an on-line scheduling system would be in place. During this time, students and advisors could access and request changes to a student's schedule electronically. The system would immediately try to schedule that request. The user would receive an immediate response: either a new schedule or an analysis explaining the rejection of the request. Up-to-date course status information would be available to system users at all times. Management query tools would be included to allow users to get answers to many different kinds of questions.

The Implementation of a Central Timetabling System in a Large British Civic University

Barry McCollum

Timetabling Officer

University of Nottingham

Nottingham, UK

Barry.McCollum@Nottingham.ac.uk

Abstract.

This paper describes the practical issues and problems which exist in relation to the initial implementation of a centralised timetabling system in an large British civic University. A chronological documentation of procedures are detailed which are associated with the particular implementation strategy adopted by the University. It concludes by questioning whether centralised timetabling is a viable option taking into consideration the culture which inherently exists in such institutions.

Introduction

The introduction of modular course structures by a large percentage of British Higher Education Institutions in the early part of this decade, has highlighted two needs namely; the provision of maximum flexibility of educational choice for students; and the optimal use of University teaching space. The central co-ordination of the University timetabling is therefore paramount if the potential gains to students offered by modularity are to be realised.

The University of Nottingham initiated a project in September 1996 with the aim of producing a central timetable encompassing the entire University by the beginning of the academic year 1998/99. The prime objectives of the project were to increase student choice within defined pathways through the modular structure while making optimum use of the available teaching space.

The intention of this paper is the following:

i. an illustration of the timetabling needs of the University of Nottingham,

ii. an overview of the progress to date regarding the implementation of a centralised University wide timetable,

iii. problems encountered to date and those foreseen in the future, and

iv. comments on a required further stage of implementation.

In addition, it demonstrates that the introduction of such a system in an old large British civic University is made complex by the long tradition of a decentralised culture.

The University of Nottingham

Following its initial establishment as University College Nottingham in 1881 the University moved to it's present site 1n 1928 and then gained full independence in 1948. The University consists of some sixty departments grouped into seven Faculties, each of which is headed by a Faculty Dean. The University is currently split into two main campuses, the main University Park Campus and the Sutton Bonington Campus. The University Park campus, which is situated close to the city centre, houses all the Faculties with the exception of the Faculty of Agricultural and Food Sciences which is situated at Sutton Bonington, some ten miles to the south of the University Park campus. The Department of Continuing Education occupies premises in the centre of Nottingham. In 1999/2000 the University intends to open a new £40 million campus, close to the University Park campus, with four academic departments relocated from University Park. The University has more than 21,600 students. Of these some 11,500 are undergraduates, 3,500 postgraduates and some 6000 follow continuing education courses.

The University's Timetabling Needs

Prior to the introduction of modularity within a semester framework at the beginning of the 1992/93 Academic Session, the University's academic regulations were established at Faculty level. This led to a lack of uniformity across the University with regard to course structure and degree regulations which discouraged the possibility of cross Faculty study. The situation was therefore one were the majority of students followed courses wholly within one Faculty. Consequentially, there was relatively little demand for cross Faculty timetabling except in a few subject areas such as psychology and geography.

With the introduction of modularisation, students were given greater control over the content of their course by allowing them to take a percentage of it as a free choice outside their main discipline. The students were therefore not confined to any one particular department or pathway for the duration of their study but were given the freedom to cross departmental and Faculty boundaries studying other disciplines according to their interest.

The major factor limiting student choice suddenly became the modular timetable. Whereas the introduction of modularity placed great effort on ensuring that course structures were made more flexible, little effort was expended on providing a basis by which this increase in flexibility could subsequently be realised through timetabling. The University's timetable therefore became fundamentally important with regard to the combination of modules which could be chosen by students. It became apparent that until a central timetable was introduced, the University would not have the opportunity of introducing novel packages within the course structure.

The courses offered pre modularisation had components significantly larger than the ten credit modules on which the new structure was built where students take 120 credits on twelve ten credit modules per annum. Hence, the complexity of timetabling increased as more events required timetabling. This complexity leads to modules being taught in rooms which were either too big or too small for the class size or did not have the right equipment. In addition, the total number of modules is some 2,400 so that the sheer amount of data made it difficult to provide timetable information in an easy assimilated form at the beginning of the academic year.

Central Timetabling

Like academic regulations, the University timetable had to be developed centrally if the flexibility offered by modularity was to be delivered. At the beginning of the current project the procedures for timetabling and room allocation varied from Faculty to Faculty within the University. This 'compartmentalised' approach to timetabling resulted in constraints being placed on the flexibility of the modular system and difficulties in introducing new courses. Therefore, the University decided that procedures relating to the timetabling of modules should be carried out centrally. In the light of this, a central Timetabling Office was established.

The recent UK National Audit Office Good Practice Guide on Space Management in Higher Education identified the following as benefits of central and computerised timetabling:

Central co-ordination of timetabling

i.   Helps deal with complaints about shortages of teaching rooms,

ii. evens out demand across the timetabled week,

iii. provides a better match between group size, room size and room availability,

iv. makes more intensive use of teaching accommodation, thereby, reducing pressures for the construction of additional space.

Computerised timetabling

i.   Copes with increased complexity, e.g. modular course structure and interdisciplinary programmes,

ii   allows modelling of different scenarios such as the introduction of new courses or the remodelling of space,

iii allows modification to accommodate late changes in course numbers and structure and student option choices.

At a time of limited funding and with the emphasis on the quality of teaching it was important that quality teaching space was used as efficiently as possible. This had been highlighted in the Pearce Report which stated that more efficient use of space will reduce the need to secure capital funding. The report also noted that greater use of computer aided timetabling and space allocation could help to improve the central allocation of space.

Implementation Structure

A Timetabling Steering Group was established including representation from all Faculties and chaired by the Senior Pro-Vice Chancellor responsible for teaching quality. This group was serviced by the author who was responsible for driving the project, setting deadlines and bringing issues of importance to the Group's attention. After the first meeting of this Group, the consultation process was started by arranging Faculty meetings with the already established timetabling officers within each department. These departmental timetabling officers had the duel role of collecting information and views from within the department and reporting back what was being discussed and how the project was to be moved forward. Information relevant at all stages of the implementation came directly from academics within all departments therefore giving as high a percentage of coverage of views as possible.

The purpose of these initial meetings was to involve staff at an early stage by consulting with them on the proposed way forward. Timetabling difficulties were discussed which in some cases were common to all Faculties and in others specific to individual Faculties and departments. In many cases discussions were held with timetable representatives on an individual basis to ensure that a full understanding was gained about the timetabling problems currently faced. It was decided on that consultation would be the single most important factor concerning the overall success of the project.

Organisational Strategy

In the development of a central timetable the University felt it important that a balance be found between student educational requirements, lecturer and departmental preferences and optimisation of usage of space. It was recognised that an overriding principle should be that, in striving towards the optimisation of space, teaching quality must not be allowed to suffer.

Diagrammatic View of overall Implementation

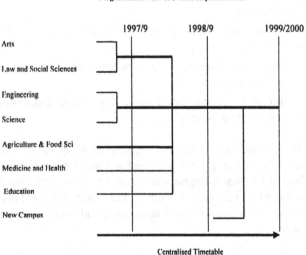

Centralised Timetable

It was agreed that a central computerised timetable would be introduced in two phases over two academic years. For the academic year 1997/98 it was decided that a central timetable be produced for the Faculties of Arts, Law and Social Sciences, Engineering and Science, and Agricultural and Food Sciences. The second phase of the implementation would take into account the Faculties of Medicine and Health Sciences and Education for the academic year 1998/99. A diagrammatic view of implementation is shown above.

Two fundamental decisions had to be made,

Should a timetable be produced before students have a choice of modules or should the students choice dictate the structure of the timetable?

To allow the student choice to dictate the timetable, even though it would guarantee maximum student flexibility, would in practice be unworkable. There would not be enough time between when the students enrolled and the timetable needed to be available. Even though the University had a system of pre-enrolment in early June, it was decided that it would be difficult to formulate a timetable on this information as many students either did not take advantage of this opportunity or changed their choices at the beginning of the following term. There was a strong argument from staff that students should be able to plan their module choice with the aid of a timetable during pre-enrolment. For these reasons it was decided that the timetable should be formulated before students have their choice of modules. It was intended that if a student felt constrained by the timetable, procedures would be put in place to allow him/her to contact the department involved to discuss the problem. If it was judged that there were good academic reasons to remove a particular module clash, the department would subsequently contact the Timetabling Office for guidance on the possibility of rescheduling that module.

Should a timetable be composed from first principles or should the current timetable be used as a basis.

Departmental timetabling officers felt that it would be preferable to construct a timetable from first principles i.e. collect information on course structure which would indicate what modules must not clash and those which could due to the impossibility or unlikelyhood of students taking them. Statistical information on what combination of modules students tended to take could be analysed from previous year's information. The alternative to this was to start the development of the following year's timetable by looking at the current year's timetable and incorporating changes which had been made in course structure or where the current timetable was unsatisfactory. It was decided to employ a combination of these strategies in dealing with the Faculties under consideration during the first phase of implementation.

Faculty Strategies

It was decided that implementation of the system for the Faculties of Arts, Law and Social Sciences, Engineering and Science would begin by modelling the current existing timetable in these areas. That is to say, producing a timetable for the Academic Year 1997/8 which is based on the previous year's timetable. New modules could be slotted into slots left by discontinued or suspended modules or they could be centrally timetabled by the Timetabling Office. The current timetable is known to exhibit many problems but it also holds extensive 'knowledge' of how individual course structures relate to each other and since time was limited it seemed more appropriate to use this rather than devise an entirely new structure from first principles. It was also agreed that this approach would substantially reduce the change in the teaching programmes for individual academic staff. In effect, module scheduling for the academic year 1997/98 would be limited to that for new modules and modules for which there were problems with the current room or time.

It was therefore agreed that the first stage in the development of a central timetable would be the collection of information on the current timetable. This was intended to identify problems with the current timetable for particular modules as well as inadequacies in room allocations. The collection of this information would make it possible to:

(i) *make informed proposals on how the timetable should be changed the following year.*

The timetable would be changed gradually in order to maximise student choice and distribute the teaching load of staff members in such a way as to free up days or half days for research.

(ii) *provide a clearer picture of how space is being used.*

This would allow modules to be re-allocated quality teaching space very quickly if it became apparent that a room was not suitable or that the number of students enrolled on a particular module exceeded the capacity of the allocated room.

Due to the structure of the courses offered within the Faculty of Medicine and Health Sciences it was decided that timetabling should be carried out as normal with information being provided to the Timetabling Office at appropriate and agreed times. This was because of the non linear structure of the timetable i.e. the timetable varies from week to week. This causes certain difficulties with the software currently being used. This along with the fact that an in-house system had recently been developed by a member of staff which seemed to be producing a comprehensive timetable led to the decision of maintaining the status quo. What was important was that detailed attention would be paid to the timetables for the courses which straddle the boundary between this Faculty and the Faculty of Science. To ensure that this process was carried out to the satisfaction of both Faculties, a group was set up comprised of timetabling representatives from both Faculties. The aim of this group was to ensure that integration is preserved from the outset.

The Timetabling Steering Group decided that as the courses offered by Faculty of Agriculture and Food Sciences had been entirely restructured for the academic year 1996/97, it was an excellent opportunity to create a timetable from first principles and use the software to schedule all modules using all the available resources.

User Concerns

Discussions with departmental timetabling officers uncovered a number of very common concerns. It was apparent that academic staff were opposed to the introduction of a central timetable due to the following;

i. *No consistency in timetable from year to year.*

This fear was allayed by taking account of the current timetable and consulting widely on where improvements could be made. The timetable produced would subsequently act as a basis for changes in subsequent years. In appropriate circumstances, as with the Faculty of Agriculture and Food Science, a timetable would be developed *ab initio*. After the initial development of a timetable subsequent timetables would be developed from previous ones.

ii. *Individuals will have no say over when or where they will teach certain modules.*

Staff saw the introduction of central timetabling as a process that actually took a certain amount of control away from them as regards to what room they taught in and at what time. Staff prefer certain rooms and times for teaching for a number of reasons which have evolved along with the course structures. As staff agreed with the principle that student choice should be increased, it was explained that as new combinations of modules become apparent the timetable would be altered accordingly in consultation with all members of staff concerned. This may mean that times at which certain modules are taught may have to change. Every effort would be made to find a suitable time for the teaching of a module to ensure that the requirements of the module are matched with the best equipped room. The aim was to meet departmental teaching preferences as much as is practicable while producing a functional timetable.

iii. *The use of the management information*

Staff were concerned that as their teaching load would be transparent this information would be used to monitor them centrally. Even though this information would be available, the majority of staff eventually accepted that the benefits the project would bring would far outweigh the negative aspects as they saw them.

iv *Decrease in student numbers.*

Staff did not want to change the weekly time modules were taught at in case less students could attend due to a clash with another part of the timetable. Staff were assured that changes would only take place after every possible clash this would subsequently create was analysed to ensure it was appropriate.

iv. *Departments would not maintain the luxury of priority over certain rooms*

This concern is dealt with in detail at a later stage.

Information Gathering

Information required was gathered by the departmental timetabling officers. Before the final timetable was produced all the following information had been collected and entered onto the software.

The information required to enable a timetable to be produced can be classified into the following:-

(i) *University*
- Dates of University teaching weeks,
- hours of teaching.

(iii) *Module*
- List of module codes and titles,
- department responsible for each module,
- requirements for each module, in terms of hours of lectures, tutorials, seminars and laboratory work,
- whether teaching takes place in single, double periods or other sized periods, and
- type & size of room module requires including special facilities required.

(iv) *Staff*
- Staff names and departments,
- which lecturers teach which modules, and
- availability of staff throughout the year.

(v) *Room*
- Size of all lecture theatres and rooms, and laboratories in terms of maximum student numbers,
- location of any specialised equipment, such as audio-visual facilities,
- location of rooms,
- suitability of rooms for particular modules,
- hours available for timetabling, and
- disability access.

Module Information

Information on modules is related to what activities are associated with a module, how often they take place and what semester the module is taught in. This was collected using a special form designed to collect all possible information on activities associated with all modules in the Faculties under consideration. As it turned out, this was an impossible task! Even though the form had been circulated to all departmental timetabling co-ordinators for comment it turned out that there was a whole host of reasons why the form was left wanting.

The overwhelming problem with the collection of this information was the accuracy of the estimates of the number of students taking modules. As timetabling for the majority of the Faculties was to be carried out before students had the chance to choose their modules, an estimate of the number of students taking modules had to be made. In many cases this field was not filled in on the form. While it was often possible to use numbers registering on individual module from the previous years as a guide, there were occasions were departments would have been able to provide more accurate information based on expected intake, numbers progressing from the previous year or the anticipated demand for a new module. Departments shied away from providing this information because they were afraid of guessing too low a number and therefore asking for too small a room for the teaching of that module. In a crude sort of logic, leaving the field blank put the onus directly on the timetabling staff to provide a suitable room at the beginning of the academic year no matter how many students registered on the module. It was also decided to build in a margin of safety in estimating the number of students registering from year to year (e.g. less than 10%) which it was hoped would result in not having to change the allocation of rooms after it was known how many students had registered. This could be done with a certain degree of certainty as it was known due to government regulations that the overall student intake of the university would not increase substantially from one year to the next. It was expected that experience over a number of years would lead to a further fine tuning in estimating the number of students taking modules in relation to the University's total intake of students.

It was also found that often the complexity displayed by certain parts of the modular structure was hard to collect. One particular example of interest came from the Faculty of Engineering where a number of individual modules were further subdivided into a number of optional courses which were common to more than one module. To illustrate this, suppose a second year module exists with number ENG 211. This module has six options A-F, which the students taking that module must choose three in the second year of their course. Now suppose a third year module ENG 311 contains the same options which students, if they choose to study it, must take the three options they did not study in their second year. When the statistical records of the last three years were analysed there seemed to be no relation to the number of people taking the module in either year to the number taking each option. Neither could this information be gathered by the form, nor could it be certain what size of rooms were required for each option.

A common problem across all the Faculties was not knowing how many tutorials were required for a particular module until the students actually turned up. In addition, in many cases there was uncertainty of who would actually teach a module due to turn over in staff or staff on sabbatical.

Staff Information

It was felt appropriate to assume that staff were available for their contracted hours unless otherwise pointed out. This placed the onus on staff members to contact the Timetabling Office to indicate when they were unavailable. For the reasons described above staff were wary of providing this information. It is expected that this will continue to be a contentious issue.

Room Information

Information on the location of rooms and their seating capacity was extracted from a database held by the Estate Office. There had already been some initial work carried out related to the collection of information on rooms during an earlier pilot of the software. This was compiled with the resulting information being loaded into the software. Problems arose relating to the equipment which was registered in each room. Often this equipment was relocated in different rooms depending on individual requirements. Because the majority of rooms were neither managed or the equipment provided centrally it proved difficult to ensure that the equipment in a room remained stationary. The potential for problems here was immense as it was difficult to timetable a module to a room that required certain equipment which might not have been available.

At the beginning of this project teaching space was managed in various ways within the University. This was generally classified into a pool of centrally booked rooms which were maintained, upgraded and equipped for teaching purposes centrally and rooms which were managed locally by individual departments who had priority over their usage. This former group of rooms were booked centrally through a central room booking service on a first come first served basis. Due to the nature of the old room booking system only minor concern was shown for matching a module to a room based on the teaching requirements of that module. The booking of the second group of rooms mentioned was carried out on a historical bases with certain departments or groups of departments having priority over the booking of certain rooms. The teaching equipment in these rooms was provided and maintained at Faculty or departmental level. If teaching was to be scheduled to these rooms centrally there had to be some extent of control over the rooms centrally to ensure that the rooms were of an acceptable standard and the correct equipment was provided. This was a major problem as Faculties and/or departments felt that they had ownership over these rooms which essentially were University teaching space.

The Management of Rooms

The full benefits of the approach to central timetabling stated above could only be realised if the Timetabling Office had the ability to allocate modules to suitable rooms. Historically, University teaching space had largely been allocated by individual departments which meant that well-equipped rooms were empty while modules were taught in inappropriate rooms elsewhere on the campus. To improve this situation details of all teaching space needed to be held centrally so that modules could be allocated the most suitable rooms for their teaching needs.

It was proposed that all teaching areas within the Faculties which are intended to be centrally timetabled for the Academic Year 1997/98 will be classified as either locally allocated or centrally allocated. In addition, centrally allocated rooms would subsequently be divided into those which were centrally managed and those which would continue to be managed locally. This classification was expected to be completed by the end of June 1997. The starting premise was that all teaching space with the exception of laboratory space will be centrally allocated, locally managed. This implied that a case would have to be made by a Faculty if a room were required to remain locally allocated or if it were proposed that a room should enter into the pool of centrally allocated, centrally managed rooms. It was stressed to the Faculties that the aim was to increase the pool of centrally bookable rooms to ensure that a greater amount of flexibility was offered during the room allocation process and their input was made welcome on what rooms in their areas, if entered into the pool, would best achieve this aim. This process was carried out in close consultation with the Estate Office. This scheme of management of rooms is shown diagramatically below.

Usage of Rooms

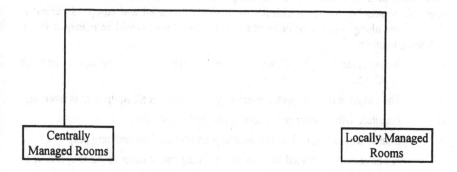

| Locally Allocated Rooms |

Departments will be responsible for allocating and managing rooms

| Centrally Allocated Rooms |

Rooms will be allocated depending on the needs of individual modules

| Centrally Managed Rooms | | Locally Managed Rooms |

- Rooms will be upgraded and maintained to a specific standard.

- Rooms will be upgraded and maintained to a specific standard by the owning department
- Money will be transferred to the department depending on how many times a room is used. This will compensate department for technicians time.

It was recognised that some rooms should remain firmly associated with particular departments and should not be available for central use, e.g. laboratories. There would also be rooms that the Timetabling Office would only allocate to classes after consultation with the department or Faculty. These might include:

i.   Laboratories which contain experimental apparatus to be used for several days teaching.

ii.   Teaching rooms kept free for impromptu research meetings or presentations by visiting lecturers.

iii.   Rooms needed for collective seminar groups.

iv.   Rooms for which external use would be inappropriate because of safety regulations.

v.   Rooms housing special or valuable collections.

Faculties were invited to identify such rooms and explain why their use needs to be restricted. Information was collected on these rooms to ensure that they remained locally bookable.

It was proposed that all other teaching space would be allocated centrally. Initially it was not intended to change existing arrangements for the local management of rooms so the department that had management responsibilities would maintain the equipment in these rooms together with an equipment inventory. Additional equipment would be available for renting at a charge to the department concerned and the purchase of new equipment would only be made after consultation with Audio Visual & Educational Services (AVES). It was decided that a mechanism would be established to ensure appropriate departments were compensated for locally managing rooms.

To increase timetable flexibility, it was a requirement to increase the number of centrally bookable rooms i.e. rooms which are maintained, upgraded and equipped centrally. These rooms along with rooms brought into this central pool would be managed in the following manner:

i.   Audio visual equipment will be provided, replaced, repaired and maintained by AVES.

ii.   The rooms will be inspected every day to ensure that all equipment is working.

iii.   Furniture will be arranged according to the first booking.

iv.   Boards will be cleaned and chalk and pens provided for teaching purposes.

v.   A notice will be placed in each room listing the number to be telephoned to report any difficulties.

vi.   A central inventory will be held of all equipment.

How and when other rooms enter the central pool would be determined in close consultation with the Estate Office, Audio Visual and Educational Services and the owning department. It was decided that after information on all rooms had been collected, a detailed report would be produced on a detailed strategy identifying which rooms should enter the central pool and when this should occur. At the time of writing this was in preparation.

Mechanism of funding

As stated above it was agreed that a mechanism would be established to ensure appropriate departments were compensated for locally managing rooms which were centrally allocated. After all these departments would be paying technicians to provide equipment for teaching which was outside their department. It was proposed that this mechanism where money was "channelled" to departments from a central source should be set up as a matter of priority. It seems apparent that rooms which would become centrally allocated and locally managed would have to be maintained to the required University standard. This implied that the owning department should receive money to ensure that this is the case. This involves the general maintenance of the room together with looking after the teaching equipment therein. It was proposed that the department of Audio Visual and Educational Services(AVES) should be consulted when departments were upgrading old or buying new Audio Visual equipment for these rooms. The part of the mechanism which posed some difficulty was deciding on how much money should be payable to each Faculty. It is proposed that after the rooms had been classified, each Faculty should be allocated an amount of money depending on the number of rooms which were classified as centrally bookable, locally managed. The amount would be based on the resources it would take AVES and the Estate Office to manage these rooms centrally. If departments made a case that they want to keep a large percentage of rooms that they had traditionally managed, they would receive little money from this mechanism of funding. This was regarded as another incentive for the departments to allow rooms to be centrally allocated while they remain under local management.

The Timetabling Process

At the time of writing this paper information on all relevant teaching space had been collected along with 90% of the required information about activities associated with modules. This information along with the amended course structure for the 1997/98 academic session is being loaded into the software. A timetable has been produced for the Faculty of Agriculture and Food Science. It was emphasized that the production of a workable solution for the first semester of 1997/98 was still in its early stages. Comments were invited on the timetable produced and amendments were asked for indicating changes in the information originally provided.

It was expected these submissions would include:-

i.   changes in the constraints that were placed on an activity. This includes issues such as changes in staff assigned to an activity or changes in the weekly teaching pattern,

ii. occasions when it is preferable for lectures to be given one after the other,

iii. new activities,

iv. explanations of why modules must be taught on a certain day. It is important to ensure that information is collated on modules which are taken by students who may have to travel to or from the main University Park campus.

It was also emphasised that the aim was to produce a workable solution under the general constraints placed on the activities and the time available in the teaching week. It may not be possible to produce a solution which is both workable and satisfactory to all.

The timetable was constructed with the aim of giving students maximum flexibility in relation to course structure. If a workable solution could not be attained, then it may be necessary to limit the choice students will have when choosing optional modules. It was pointed out that this is for the Faculty of Agriculture and Food Science to decide upon. It may be the case that 'tactical clashes' need to be established in the timetable. At the time of writing, activities associated with twelve modules were not able to be timetabled. It was intended that the second stage of timetabling will reduce this number substantially.

Future Issues

There are essentially two methods for the future production of a University timetable. That is, centralised or decentralised timetabling. The centralised approach adopted by the University of Nottingham involves having a single installation of the timetabling software on a PC workstation. All data input, timetabling, and output, is done on one workstation by a single timetabler.

If the work of timetabling is to be shared, then more than one installation of the Timetabling software is necessary. In this case, the timetabling data must be consistent between the installations. To ensure this consistency, the timetablers must use a common data source. This might be a file on a server to which both PCs connect; however this is not a very secure solution as files can be corrupted if the software used crashes, or the network connection is lost while updates are being made. It may also be unclear what might happen if simultaneous updates to the data file are attempted from more than one workstation.

The software chosen offers the 'Scheduling Database' as an option. The timetabling data is held on a central server computer, and is down loaded to PC workstations when needed. Users can be given permission to modify particular parts of the data, or only to view it. Network faults and PC malfunctions should not corrupt the centrally stored data, although updates may be lost. The software would manage any attempts at simultaneous updates in a well defined manner.

In an age where the emphasis is in the dissemination of information electronically it might seem as though the University is breaking the trend centralising the procedures related to the production of a timetable. However, the dissemination of the timetable information in electronic form around campus is an important part of the timetabling process. A number of installations could be envisaged of the software around campus, used simply to view the timetabling data, or perhaps to do some local modifications such as timetabling locally controlled rooms. These installations could be set up so they are unable to change any of the centrally managed timetable. It is envisaged that perhaps this process could be carried via a web server. The alternative solution to disseminating information to desktop PCs is to write our own interface to the timetabling software, perhaps writing HTML for WWW browsers to read. This would avoid the need for distributed users to learn the software.

The author is also keen to keep abreast with the developments of the in-house software, Automatic Scheduling and Planning (ASAP). The development of this software is funded by the Joint Information and Systems Committee of the British Higher Education Funding Councils.

New Campus

The development of the new campus will pose the following issues in relation to timetabling.

i.   students will have to be allowed adequate travelling time between the campuses,
ii. the possibility of integrating the use of video conferencing between campuses into the timetable,

Difficulties that arise during the implementation phase will be taken into account in the development of future timetables.

Campus zoning would be introduced to ensure minimum disruption to students moving between lectures. This is not currently available within the software being used.

Conclusion

A description has been given of why timetabling is crucial in meeting the Universities aims and objectives with regard to student choice, what the procedures and problems have been to date with the implementation process and a look ahead to the differing strategies of timetabling that could possible be employed post the current implementation phases.

It is proposed that there are two fundamental issues in realising the aims of the project currently under discussion. The first is the institutional management culture and the second the territorialism of the staff. It is argued that the latter is a direct consequence of the first.

The old British civic Universities are characterised by having a strongly defined departmental structure with many of the procedures intrinsic to the running of the institution operating on this basis. Departments control their own budgets and to a certain extent have their own autonomy with respect to decision making and Faculty procedures. This structure of management has resulted in individual departments and sometimes groups of departments having their own procedures which may or may not be similar to the rest of the institution. These procedures have been allowed to evolve with central management concerned with the overall performance of the University and operating a policy of devolution to local management in departments. It is this non standardisation in procedures that makes it difficult to try and implement any centralised system across the entire University whilst paying heed to everybody's needs. This situation is in contrast to Higher Education institutions established in the 1960's which have a much stronger central culture. Even though departments exist, there is a strong central lead over institutional and departmental procedures. is managed. These institutions do not have the hurdle persuading staff the value of common procedures.

The issue of implementing a centralised University software system is therefore a management issue as much as a software issue. Procedures must be established that break down the information barriers which are a direct consequence of the existing culture within the institution. The aim of the project is to facilitate both student and staff needs while making effective use of quality teaching space. It is proposed that, within the current structure, this will be achieved by the dissemination of timetabling information to departmental level, allowing them the possibility to carry out their own modelling of course structure and create their own timetables. This would therefore give departments the control over the rooms they so dearly want to retain under local management. It is envisaged that this would be co-ordinated centrally as opposed to formulated centrally. The process would therefore have evolved from centralising the production of a timetable and standardising the associated procedures to gradually providing departments with access to the information and hence giving them the opportunity to carry out their own timetabling within the limits of the overall modular structure. Another management issue in the debate is the extent to which academic departments have dedicated administrative support. Many departments in Nottingham are not large enough to have full-time administrative support so that devolution to departments means that the tasks are carried out by academics thereby reducing their time for teaching and research. If tasks are then delegated to different staff within the department an extra tier of co-ordination is required. For example, the information about the modules being taught as provided to the timetabling office by the departmental timetabling officer might not be the same as that provided to a different part of the administration which prints the module catalogue. A single source of information is required within departments to eliminate these problems.

Fundamentally, timetabling is arranging a defined number of events in both time and space under specific known constraints. At this first stage of implementing a centralised University timetable the problem of non standard procedures across departments has been addressed. The perceived flexibility offered by central timetabling will not materialise until the central system has the scope to allocate events to a room based on the needs of that event. This culture of territorialism therefore needs to be eradicated. The strategy for doing this is gradually chipping away at it using the method detailed earlier by centrally allocating all rooms and providing money for those rooms which are not centrally managed

The pressures placed upon timetabling by the increasing demands of a modular structure is fast making institutions realise that space management is a priority if courses are to be delivered in a manner which will provide flexibility for student choice and quality teaching space for those courses to be delivered. In addition to these methods for increasing the number of centrally booked rooms, it is proposed there must be an awareness fostered of the cost of space. The culture which sees space as a free good must also be eroded. The Estates Office must be accepted as a service to the institution. The concept of space charging as an effective space management technique is not new with other higher institutions having already implemented this strategy. The first stage in the implementation of this would be the development of a methodology of space charging. This would include issues such as definition of quality space, how much should be charged, how this will be monitored and collected etc. The University of Nottingham are currently of the frame of mind that the amount of administration required to develop and implement a methodology required for Space Charging is too great relative to the likely benefits.

At the time of writing, the project described is very much in its early stages. The emphasis of the description here is placed on the initial problems in relation to the acceptance of a centrally controlled University timetable. It is expected that during the next phase of implementation more problems will arise which are directly related to the timetabling process.

A Brute Force and Heuristics Approach to Tertiary Timetabling

T. Nepal[1], S.W. Melville[2], and M.I. Ally[1]

[1] ML Sultan Technikon, Centenary Road, Durban 4001 South Africa
e-mail: nepal@wpo.mlsultan.ac.za
[2] University of Natal, Durban 4041 South Africa

Abstract. This paper describes a study of existing tertiary timetabling techniques in Southern Africa, based on initial questionnaires and follow-up interviews. Heuristics obtained from this study were used as the basis for a prototype automated timetabling package. The prototype's design and implementation is discussed. Results obtained using the prototype on both a battery of prepared tests as well as a real-life timetabling problem are discussed. The performance of the prototype model gives strong indications that the final system will be a competent timetabler.

1 Introduction

A number of approaches have been employed to tackle the timetabling problem. These approaches have been widely reported: Graph Theoretical approaches ([12,27]), Operational Research approaches ([5]), Boolean Matrix Iterational approaches ([2,19]), Expert System approaches ([16,24]), Simulation approaches including Genetic Algorithms ([9,1,21,10,23]), Simulated Annealing ([1,14]). Mathematical programming approaches include Exchange procedures ([15]), Lagrangean relaxation ([7,30]), and Tabu Search ([8,6,13]).

The following issues have been highlighted and need to be considered: The rooms problem ([22,18,29,31,26]), unavailabilities ([4,12,11,28]), minimising versus equitation ([17,22,3]), preferences ([17,25,26]).

Both Lewis [20] and Punter [27] stress that there is a vast number of requirements specific to institutions and courses and that generalised solutions must be adapted to individual circumstances. At technikons, the timetabling problem involves scheduling lecturers, subjects, groups, and rooms to a number of periods of a week.

Technikons offer diplomas referred to as courses e.g., National Diploma in Information Technology. The courses are divided into 1st, 2nd, and 3rd levels of study. Each level at the technikon is regarded as a group e.g., 1st, 2nd, 3rd year group. Timetables are prepared for each group and not each subject. Common subjects offered by more than one diploma are typically combined.

2 Empirical Research

Apart from doing a literature survey on prior research, empirical research was carried out by the authors on timetabling techniques in southern African tertiary institutions. Questionnaires were designed and sent to timetable co-ordinators of the different tertiary institutions. A total of 46 questionnaires were despatched. There was a return rate of over 80% in the case of Universities and Technikons. Thirty-six institutions indicated that the timetable was constructed manually, fifteen used a mixed approach using either the trial and error method or the so called "blocking system".

Follow up interviews with timetable co-ordinators of 3 technikons and 3 universities were conducted. In all institutions the master timetables were constructed manually. In most cases the timetabling was decentralised to faculties or to those levels at which the timetabling problems are largely independent of one another. Undergraduate lectures were generally timetabled at faculty level; tutorials and postgraduate teaching at the departmental level. Service classes, provided by one faculty for other faculties, require co-ordination and are in general timetabled first so that the remainder of the problems can be dealt with independently of one another.

The findings of the empirical research and that of the interviews were used in formulating heuristics that can be used in an automated timetabler. These heuristics are discussed in the next section.

3 Heuristics

One heuristic that arose in all interviews could be generalised to the rule "*Schedule scarcest resources first*". In all cases the specific resource that was scarce was venues, particularly large venues.

Another important issue to all respondents was that of *equitation*. Having periods for a given subject spread evenly across the week was clearly highly desirable or even vital to respondents. Didactically the justification for this seems obvious: having all 6 periods, say, of a given subject all on the same day would hardly be conducive to a positive teaching/learning environment. This gives us our second heuristic, "*Spread the periods for a subject as evenly as possible across the week*".

A third point to arise from the interviews is less easy to didactically justify, but nonetheless is definitely important to tertiary timetablers. This is the idea that periods for a subject should be "mixed" across the week. In other words a situation where a given subject required five sessions a week and all five of these sessions were in the third period, say, was unacceptable. Timetablers interviewed unanimously felt that every subject should have a mix of "bad" periods (e.g. early morning, late afternoon) and "good" periods (e.g. mid-morning) rather than some subjects having all the desirable periods and others all the undesirable ones. This leads to our third heuristic, "*Ensure that subjects get a mix of assigned periods*".

A simple yet effective heuristic suggested by Blakesley [4] is to treat unavailabilities (of lecturer, class or venue) as preallocations to "dummy" resources (i.e. if it is known that lecturer C is unavailable on 10th period Tuesday, then we create a preallocation for 10th period Tuesday containing lecturer C together with class X and venue X, where the class and venue are non-existent dummy entries). Thus both unavailabilities and preallocations can be treated as the same type of problem rather than two separate ones. Our heuristic here is "*Convert unavailabilities into preallocations*".

As preallocations (and unavailabilities converted into preallocations) must necessarily occupy predefined periods, it is a common-sense heuristic to schedule these before any other scheduling takes place. This averts the situation where one of the three key resources (lecturer, class and venue) involved in the preallocation gets provisionally "booked" in the needed period prior to the preallocation being "booked". This leads to the heuristic "*Schedule preallocations before anything else*".

Combining or splitting groups leads to its own set of problems, as evidenced by both our questionnaire and our interview results. However, it is often desirable to combine groups into a large class for a common subject, and similarly to split a large group into small sub-groups for, for example, tutorials. A suggested heuristic here was "*Treat sub-groups of a split class as individual scheduling entities for purposes of the split, while treating them as a combined class for those times when they are not split*". This has the advantage that we then need only consider the special case of a combined class rather than the additional case of a split class—a split class is treated as a number of unique classes. In addition it allows a great deal of flexibility in the resultant timetabler—six classes of varying sizes could be combined for a given subject, and split into twelve distinct groups of equal size for tutorials.

4 Prototype Automated Timetabler (PAT) Design

4.1 Implementation Languages

The prototype's user interface is written in Clipper, while its actual scheduling code is in Pascal. The Clipper program (TTMAIN) does the front end screen capture of the necessary data and prepares the data files for input to the Pascal program (TTGUTS) which does the scheduling. The printing of timetables is handled from the Clipper program.

Clipper is a flexible base language in which one can write efficient, encapsulated, and reusable program modules. User interfaces can be programmed in a simple and efficient way. It is therefore suitable for prototyping compared to third-generation languages. Clipper is therefore an appropriate 4GL for user interface and for input and output screens and forms generation. However for the actual scheduling of a timetable requires considerable array manipulation, formulae usage, and some randomisation. Clipper does not lend itself to such computation and the two languages that provide such facilities in a programmer-friendly form together with acceptable file I/O for interfacing with Clipper data are the 3 GLs, C and Pascal. Pascal was chosen in preference to C due mostly to the authors' familiarity with the former language.

4.2 Overview of the Timetabler

Inputs. Inputs to the timetabler entered through the Clipper front-end are as follows:

- *Lecturer Data* – a list of all lecturers in the institution. A unique code is generated for each new entry to create a lecturer description (name) and code file.
- *Venue Data* – a list of all available venues. In addition to a venue description the user is prompted for venue type (e.g. normal, computer room, studio) and venue capacity. Again unique codes are generated and all information stored in an appropriate file.
- *Subject Data* – a list of all subjects offered at the institution. In addition to the subject name, the number of periods required per week for the subject must also be provided.
- *Group Data* – a list of the various class groups. The user is prompted for diploma name (e.g. Human Resources Diploma), year of study, and class size.
- *Group/Subject Data* – a list of what subjects are taken by each group. In addition to entering this data the user must also indicate if she/he wants classes combined (or split, for that matter) for a particular subject—if this is the case an appropriate flag is set in the resultant file.
- *Lecturer/Subject Data* – a list of what lecturers will be lecturing which subjects.
- *Preallocation Data* – a list of preallocated periods (for class, venue and lecturer). The user may also enter an unavailability here; this is treated as a preallocation in storage (transparent to the user) but with flags for the two resources not being reserved to indicate that these are dummy values.

Processing. Processing begins by constructing a Master Table of records showing entities needing to be scheduled, based on the user-supplied data. Each record in the table contains the following fields:

Lec   The lecturer involved (in the case of multiple lecturers for a single subject a separate record will be created for each lecturer).

Subj   The subject involved.

Num_Stud   The total number of students taking this subject. (This could be a summation in the case of a combination—in the case of splits each split will be represented in a separate record.)

Num_Pers   Number of periods required for the subject.

Grps   An array holding the group code for each group (class) taking the subject—in the case of combinations there would be more than on group.

Num_Grps   Holds a value indicating the number of groups involved.

In addition to creating this Master Table, vectors of unavailability are created according to the user input preallocations. For each group, lecturer and venue an "unavailability vector" is constructed defining those periods in which the resource may not be used. Splits and combinations are also dealt with at this stage to enable proper construction of Master Table records, with each split group forming a separate record.

The main driver in the software is Procedure PROCESS, which reads as follows:

```
PROCEDURE PROCESS( VAR GOT_A_VALID_TIMETABLE : BOOLEAN);
    VAR SOLVED, SOLUBLE: BOOLEAN;
                ORDERING: INTEGER;
BEGIN
    START_TIMER;      {Start timing run}
    SOLVED:= FALSE;
    SOLUBLE:= FEASIBLE;      {Check no blatant impossibilities
                                    with function FEASIBLE}
    EQUITATION:= 100;
    WHILE ((EQUITATION >= 0) AND SOLUBLE AND (NOT SOLVED)) DO BEGIN
        REPEAT
            GET_NEXT_ORDERING( ORDERING);
            ATTEMPT_ALLOCATION( SOLVED);
          UNTIL (SOLVED OR TRIED_ALL( ORDERING));
        IF (NOT SOLVED) THEN EQUITATION:= EQUITATION - 50;
    END;
    GOT_A_VALID_TIMETABLE:= SOLVED;
    END_TIMER;      {End timing run}
END;
```

This is the main driver routine of Timetable allocation. It starts with a test to function Feasible, which checks that the input data theoretically allows for a valid timetable (e.g. that no lecturer has a lecture load greater than the number of periods in a week).

A **WHILE** loop is then entered which will run until either a solution is found or it is determined that no solution is possible. Within this loop all orderings (innermost) of allocations are attempted within different equitation levels. (Clearly we begin with equitation at maximum, and decrement only when it is found that no solution is possible at that level.) Equitation affects the "legal" number of periods for a particular lecturer-subject-group combination on a day. At equitation $= 100$, for example, a 6 period per week course would be allowed exactly one day with 2 periods, and 4 days with one period; and with equitation at 0 all six periods could be allocated on the same day.

The innermost loop in process calls two routines—Get_Next_Ordering and Attempt_Allocation. Get_Next_Ordering is currently a dummy routine which will be implemented in the final version. The idea of the routine is to progress through all possible sequences of period allocation to give every chance for a solution—currently however all test examples (from low to high complexity and varying capacity use) are correctly allocated without needing this. Attempt_Allocation is a key routine which is described below.

Attempt Allocation. This routine lies at the heart of the program. It begins by calling Set_Up_Bookings, which initialises bookings for venues, lecturers and groups (2D arrays indexed by subject (e.g. lecturers) and period) to Empty, and then sets those entries corresponding to preallocations and unavailabilities earlier read in to Booked.

A loop is then entered where Master Records (from the Master Table described earlier) are then taken one by one and allocated. The first step in this loop is to call Order_Legal_Venues, which will return a list (OK_List) of venues whose capacity equals or exceeds the group size specified in the master record being handled. The list is sorted with the "tightest fits" appearing first. (E.g. If we had 4 venues of size 20, 50, 60 and 100, and the master record group size was 45, then OK_List will return 50, 60, 100.) This is in line with the heuristic to maximise efficient resource use—try the best fit first.

If OK_List is non-empty (empty signals a failed attempt), then the procedure Get_Maxs_On_Equitation is called which does some arithmetic to determine, for the number of periods required for the subject (specified in the Master_Rec involved), what the maximum number of periods per day are and for how many days that maximum may be used, given the current equitation level. E.g., For equitation = 100 and 7 periods needed, Max_Per_Day would be set to 2 and Days_At_Max also to 2 (2 days with 2 periods and 3 with 1). At equitation = 50, Max_Per_Day would remain 2, but Days_At_Max is set to 5 (so could have 2 periods on 3 days and just 1 in the other 2 together, for example). At equitation = 0, Max_Per_Day is set to total number of periods for the week.

The order of days in the week is then randomised by Set_Random_Order; this prevents for example, all six-period subjects choosing Monday as first choice for their day when 2 periods are needed. Now a loop is entered which runs while periods are left to allocate (periods initially set to the number required as indicated in the master record, and decremented by one on each allocation) and while allocations are possible (if no allocation is possible we must try again with different equitation). An inner loop runs from 1 to Num_Days (set at 5 at present, this constant can be altered if a different scheduling period—e.g. fortnight—is required).

If Equitation_Ok (the maximum periods for the day not yet utilised) and Space_On_Day (at least one unbooked period exists in the OK_List of venues for that day), then Fill_A_Slot is called to perform the actual allocation.

Fill_A_Slot begins by randomly assigning period preferences. (In a non-prototype version, preferences will be entered by the user; for now this prevents overuse of early periods and ensures a spread of periods in line with the earlier identified heuristic.) It then repeatedly takes available periods from the OK_List of venues, trying all periods available in the best venue, then the next best and so on, until an allocation can be made or it is found that no allocation is possible. Fill_A_Slot uses Can_Slot to test, given a venue and a period of a day, that both lecturer and group for the subject are free. If so then Can_Slot sets the venue, lecturer and group booking arrays to Booked in that period, and returns success, else the next period is tried.

Outputs. Final allocations are passed via file from the Pascal program to the Clipper interface. This latter allows the user to specify required output format(s) (e.g. lecturer timetables, class timetables, venue allocations).

5 Results

5.1 Test Data

To test the performance of the prototype in a rigorous fashion, 67 contrived test cases were presented, as well as real-life data. The contrived test cases were chosen in order to present the prototype with ever more difficult changes.

Contrived Test Data. In all cases 160 subjects must be scheduled and 5 periods is set as the required number of periods per week for each subject, so ensuring a total of 800 periods to be timetabled. This means that in a 5 day, 10 period per day week, the minimum feasible number of venues needed is 16 (800/50), with all venues utilised to maximum capacity.

We begin testing with scenario 1, where each lecturer lectures only a single subject and vice versa. 50 venues are provided, all of which have capacities exceeding the maximum class (group) size. Each group takes 4 subjects.

In scenario 2 to 8 we keep the single subject/single lecturer constraint, but provide venues of varying sizes, so testing that the timetable can handle appropriate allocations of venues to classes. In these scenarios the number of available venues is gradually decreased from 50 (scenario 2) to 16 (scenario 8—maximum capacity).

Scenario 9 to 57 are sets of 7 tests, as 2 to 8 were, again with the available venues being decreased from the first scenario in the set (lots of spare capacity) to the seventh and last in the set (100% utilisation). What is varied from set to set is the loading on the lecturers, which is steadily increased. By scenario 57 we are looking at a fairly extreme loading on lecturers (40 periods each per week) as well as well as venues (100% utilisation needed for the timetable to work).

Scenario 58 to 67 allow adjustment on group loadings to be done in tandem with lecturer and venue loading adjustments. In scenario 67 we have groups taking 9 subjects (45 periods per 50 period week), lecturers taking 8 subjects (40 periods per week) and 17 venues (94% capacity).

It was felt that if the prototype could handle cases this extreme, there would be strong evidence that an automated timetabler would be practical. In addition the gradation of test cases (easier to harder) would hopefully allow for useful performance analysis.

Real-life Test Data. Given the performance of the prototype on the tests presented, it was decided to attempt to run it on a real-life problem. The ML Sultan Technikon's Commerce Faculty timetable for 1997 was selected as a useful real-world problem: devising a timetable for this had just taken up some 3 weeks (full-time) of a committee of six academics.

Some particulars of the timetable needed are given below:

Total Students	2065
Total Lecturers	31
Total Groups	35
Total Subjects	35
Subjects per group	Varying from one to six. (Groups doing only one subject are from other faculties.)
Subjects per lecturer	Varying from one to four.
Group sizes	Varying from 15 to 120.
Largest combination	Principles of Information Systems, where several groups are combined giving a class size of 720, which is split into three groups being lectured by three different lecturers.
Venues	2 × 300-seaters, 4 × 120-seaters, 2 × 90-seaters, 2 × 80-seaters, 3 × 60-seaters, 1 × 50-seater, 3 × 45-seaters, 2 × 25-seaters.

The PAT system took 147 seconds to produce a valid timetable for this problem, with 100 weight on equitation (i.e. even spread of all subjects' periods throughout the week). Further, it needed to use only nine of the 18 venues actually available to the Faculty, and with no periods scheduled for these venues they are therefore free for other uses.

Contrasting the results of the automated timetabler against that of its human counterparts is not entirely fair, as the human timetablers did pay some consideration to lecturer preferences, something not catered for in the computer system. However, it is difficult not to make the comparison of the 2 minutes of computer time against the 18 person-weeks of human time.

5.2   Analysis of Time Performance

The prototype solved all timetabling problems in times that would be very acceptable in real world usage. Using a Pentium 100, in all scenarios bar one the time taken was under three seconds, whilst scenario 8 took just over a minute. (Quite why this seemingly easy case consistently takes longer is still a subject of debate.) Spending less than a minute to construct timetables involving, e.g., 40 separate class groups, comprising 3 000 students, 80 lecturers each taking 2 subjects for 160 subjects in total, all crammed into a 100% utilised 16 venues (scenario 15), has considerable benefits over the manual alternative. The real-life test case provided further evidence in this regard.

Analysis of how performance degrades as problem complexity increases (fewer and fewer venues, greater and greater loading on lecturers and groups) was expected to be a simple matter of statistics. However, no statistically significant degradation occurred as the problem become more complex, and indeed, times seemed much the same throughout.

5.3 Analysis of Accuracy

The timetabler consistently produced valid timetables with periods spread equitably over days in the week, and subjects having a good spread of periods of the day. All resultant timetables were manually cross-checked and found to be valid.

6 Conclusion

After an initial literature study, this research began with a questionnaire-based study of timetabling in Southern African tertiary institutions. It was apparent from the responses that there was little automation of the timetabling process in these institutions, and further that a wide variety of different approaches were being used. Importantly, the process was consuming a considerable amount of personpower (several weeks of academics' time each year in some cases).

Follow-up interviews were conducted with individuals responsible for timetabling at major South African institutions. From the questionnaire and interview responses a number of heuristics used by "domain experts" were discovered.

The heuristics were embedded in code to form a prototype automated timetabler. The system was intended to be purely a test system for "toy" problems, and was based on applying these heuristics to guide an otherwise brute-force exhaustive search technique for allocations.

A fairly intense battery of tests were presented to the prototype, with problems varying in complexity from simple to extremely difficult (where resources were used at 100% of capacity). In addition a major real-life timetabling problem was presented as a test case. Against expectations, the prototype had no difficulty in producing valid timetables for all test cases, and producing them in very little time (a worst case of 147 seconds for the real-life example).

It seems clear that the "brute force and heuristics" approach, while perhaps not terribly elegant, holds enormous potential as the basis of a full-blown automated timetabler.

Future work will concentrate on the development of such a full-blown system. This system will be designed to handle some of the desirable attributes for a timetable which were omitted for purposes of the prototype model. This would include the following:

- *Catering for lecturer/class preferences.* This could be handled in a similar manner to that in which equitation is currently handled—begin by attempting allocations with preferences being set at maximum weight, and gradually decrease the importance of such preferences until a valid timetable is produced. Unlike equitation, which did not seem to be a problem for the prototype at least (maximum wight equitation was achieved in all cases), the fact that a number of class/lecturers are likely to have similar preferences does indicate that achieving timetables with all preferences satisfied is unlikely.

- *Catering for required double or triple periods.* The prototype can currently accept any number of periods for a subject, but there is no mechanism for the user to indicate that certain periods should be run consecutively.
- *Generalising the use of specialist venues.* In the prototype, the only way to enforce the use of a specialist venue at certain times by certain groups is to make these preallocations. All venues are considered the same (apart from differences in capacity) in the normal allocation process. The full version will allow special venues (e.g. computer laboratories) to be specified as such and for information regarding a subject's need for these to be captured from the user. This will allow these venues to be handled in a more transparent and far simpler way for the user than the preallocation route.
- *Catering for team-teaching.* The prototype will allow classes to be split easily enough, but does this by the simple expedient of dividing the class size by the number of lecturers for the subject, and then allocating each lecturer her/his "portion" of the split class as a new class. In practice numerous other techniques may be combined when lecturers share a subject—different lecturers may take different time slices (e.g. first 6 weeks with lecturer A, second 7 weeks with lecturer B) or the lecturers might each take some of the subject's lecture periods each week. Catering for these various approaches (and making the timetabler flexible enough so that it can accept any desired approach) will no doubt be a fairly arduous task, however on the basis of experience with the prototype it nonetheless seems a very achievable one.

References

[1] Abramson, D.: Constructing School Timetables Using Simulated Annealing. Manage. Sci. **37** (1991) 98–113.

[2] Aust, R.J.: An Improvement Algorithm for School Timetabling. Comp. J. **19** (1975) 339–343.

[3] Bardadym, V.A.: Computer-Aided School And University Timetabling—The New Wave. In: Proceedings of the 1st International Conference on the Practice and Theory of Automated Timetabling (1995) 253–268.

[4] Blakesley, J.K.: Automation in College Management. College and University Business **27** (1959) 39–44.

[5] Carlson, R.C., Nemhauser: Scheduling to Minimize Interaction Cost. Oper. Res. **14** (1965) 52–58.

[6] Carmusciano, F.: A Simulated Annealing with Tabu List Algorithm for the School Timetable Problem. In: Proceedings of the 1st International Conference on the Practice and Theory of Automated Timetabling (1995) 231–243.

[7] Carter, M.W.: A Lagrangian Relaxation Approach to the Classroom Assignment Problem. INFOR. **27** (1989) 230–246.

[8] Chan, H.W., Sheung, J.: Roster Scheduling at an Air Cargo Terminal: A Tabu Search Approach. In: Proceedings of the 1st International Conference on the Practice and Theory of Automated Timetabling (1995) 409–422.

[9] Colorni, A., Dorigo, M., Maniezzo, V.: Genetic Algorithms and Highly Constrained Problems: The Time- Table Case. (1990)

[10] Corne, D., Ross, P., Fang. H.-L.: Fast Practical Evolutionary Timetabling. In: Proceedings of the AISB Workshop on Evolutionary Computation, Springer Verlag, (1994).

[11] Day, D.P.: A Graph-Theoretical Approach to Timetabling. Unpublished MSc thesis: University of Natal, 1988.

[12] De Werra, D.: An Introduction to Timetabling. Europ. J. Oper. Res. **19** (1985) 151–162.

[13] De Werra, D.: Some Combinatorial Models for Course Scheduling. In: Proceedings of the 1st International Conference on the Practice and Theory of Automated Timetabling (1995) 1–20.

[14] Dowlands, K.: A Timetabling Problem in which Clashes are Inevitable. J. Oper. Res. **41** (1990) 907–918.

[15] Ferland, J.A., Lavoie, A.: Exchange Procedures for Timetable Problems. Discrete Appl. Math. **35** (1992) 237–253.

[16] Gudes, E., Kuflik, T., Meisels, A.: On Resource Allocation by an Expert System, Eng. Appli. Art. Intel. **3** (1990) 101–109.

[17] Konya, I., Somogyi, P., Szabados.: A Method of Time-Table Construction by Computer (1978) 171–181.

[18] Laporte, G., Desroches, S.: The Problem Of Assigning Students to Course Sections in a Large Engineering School. Comput. Ops. Res. **13** (1986) 387–394.

[19] Lazak, D.: A Heuristic Approach to Algorithmic Intended for the Solution of the Timetable-Problem. Computing **4** (1969) 359–367.

[20] Lewis, C.F.: The School Time-table. Cambridge: The University Press (1961).

[21] Ling, S.-E.: Integrating Genetic Algorithms with a Prolog Assignment Program as a Hybrid Solution For a Polytechnic Timetable Problem. In: Parallel Problem Solving from Nature 2, Elsies Science Publishers, B. V., Manner & Manderick (1992) 321–329.

[22] Loo, E.H., Goh, T.N., Ong, H.L.: A Heuristic Approach to Scheduling University Timetables. Comput. Educ. **10** (1986) 379–388.

[23] Luchian, H., Ungureanasu, C., Paecher, B., Petriuc, M.: Fine-tuning a Genetic Algorithm for the Timetable Problem. In: Proceedings of the 1st International Conference on the Practice and Theory of Automated Timetabling (1995) 435–442.

[24] Martinsons, M.G., Cheong, K., Chow, P.K.O., Lo, V.H.Y.: Academic Timetabling With a Microcomputer-Based Expert Systems Shell. Artif. Intel. Educ. **4** (1993) 333–356.

[25] Monfroglio, A.: Timetabling Through a Deductive Database: A Case Study. Data Know. Eng. **3** (1988) 1–27.

[26] Nepal, T.: An Investigation into Timetabling at Tertiary Institutions in Southern Africa. In: Proceedings of the South African Institute for Computer Scientists and Information Technologists (1995) 285–292.

[27] Punter, A., School Timetabling by Computer I: A Graph Colouring Formulation. (1977) 23–25.

[28] Rankin, R.C., Memetic Timetabling in Practice. In: Proceedings of the 1st International Conference on the Practice and Theory of Automated Timetabling (1995) 45–56.

[29] Selim, S.M., An Algorithm for Constructing a University Faculty Timetable. Comput. Educ. **6** (1982) 323–332.

[30] Tripathy, A., School Timetabling—A Case in Large Binary Integer Linear Programming. Manage. Sci. **30** (1984) 1473–1489.

[31] White, G.M., Wong, K.S.: Interactive Timetabling in Universities, Comput. Educ. **12** (1988) 521–529.

Other Timetabling Presentations

Other Timetabling Presentations

Other conference presentations are listed below together with the addresses of the authors.

Title: Weekly Lecture Timetabling with Genetic Algorithms
Authors: P. Adamis and P. Arapakis
Address: Department of Informatics, Technological Educational Institution of Thesaloniki, Greece 541 01.

Title: Heuristic Graph Coloring For Consecutive Lectures Constraint in Timetabling Problem
Authors: JongIl Ahn and TaeChoong Chung
Address: Department of Computer Engineering, KyungHee University, YouInSi, KyungKiDo 449-701, Korea.

Title: University Self-Registration And Automatic Section Assignment
Authors: R. Alvarez-Valdes, Eric Crespo and Jose M. Tamarit
Addresses: R. Alvarez-Valde, J.M. Tamarit, Department of Statistics and Operations Research, Facultat de Matemàtiques, Universitat de València, Dr. Moliner, 50, Burjassot València, Spain.
E. Crespo, Department of Financial and Mathematical Economy, University of València, València, Spain.

Title: The Syllabus Plus Exam Scheduler: Design Overview and Initial Results
Authors: R. Barber and G. Forster
Address: Scientia Limited, Cambridge, England.

Title: The ASAP Timetabling System: User Interface and Problem Specification
Authors: E.K. Burke and K. Jackson
Address: Department of Computer Science, University of Nottingham, University Park, Nottingham, NG7 2RD,UK.

Title: Introducing Non-determinism into Heuristic Based Algorithms: An Investigation
Authors: E.K. Burke, J.P. Newall and R.F. Weare
Address: E.K. Burke, J.P. Newall,Department of Computer Science, University of Nottingham, University Park, Nottingham, NG7 2RD,UK.
R.F. Weare, Cap Gemini UK plc, 51 Grey Street,Newcastle, NE1 6EE,UK

Title: School Timetabling using Genetic Search
Authors: J.P. Caldeira, A.C. Rosa
Address: Laseeb - ISR - IST, Portugal.

Title: EXAMINE: A General Examination Timetabling System
Author: M. Carter
Address: Mechanical and Industrial Engineering, 5 King's College Rd, University of Toronto, Toronto, Canada

Title: Comparison of two approaches for complex employee scheduling
Authors: Y. Caseau and T. Kökény
Address: Bouygues - Direction Scientifique, 1 avenue Eugéne Freyssinet, 78061 St. Quentin en Yvelines cedex, France

Title: Practical School Timetabling: A Hybrid Approach Using Solution Synthesis And Iterative Repair
Authors: H.W.Chan, C.K. Lau and J. Sheung
Address : Department of Computing, Hong Kong Polytechnic University, Hong Kong

Title: Automatic Class Timetabler (ACT) for University
Authors: TaeChoongChung
Address: Department of Computer Engineering, KyungHee University, YouInSi, KyungKiDo 449-701, Korea.

Title: Tradeoffs and Phase Transitions in the interactions between exam-splitting, clash, near-clash, and room capacity constraints
Author: D. Corne
Address: Parallel, Emergent and Distributed Architectures Laboratory, Department of Computer Science, University of Reading, Reading, UK.

Title: A Test Bench for Rostering Problems
Author: P. De Causmaecker, G. Vanden Berghe, A, Heeffer and A. De Witte
Addresses: P. De Causmaeker and G. Vanden Berghe, KaHo St.-Lieven, Dept. KIHO, Gebroeders Desmetstraat 1, B-9000 Gent, Belgium.
A.Heefer and A. De Witte, Impakt N.V. Ham 64, B-9000 Gent, Belgium.

Title: Timetabling with Neeps and Tatties
Author: A. Cumming
Address: Department of Computer Studies, Napier University, Edinburgh, UK.

Title: High School Timetabling in Germany - Can it be done with MIP?
Authors: H. Hilbert
Address: Stephanstr.62, 10559 Berlin, Germany.

Title: A Generalized Linear Programming Model for Nurse Scheduling
Authors: B. Jaumard, F. Semet and T. Vovor
Addresses: B. Jaumard and T. Vovor, Département de mathématiques et de génie industriel, GERAD and École Polytechnique de Montréal, C.P. 6079, succursale Centre-ville, Montréal (Québec), H3C 3A7, Canada
F. Semet, Département d'Administration de la Santé, Université de Montréal. C.P. 6079, succursale Centre-ville, Montréal, (Québec),H3C 3A7, Canada.

Title: Minimax Approaches to the Faculty-Course Assignment Problem
Authors: I. Kara and M. S. Oxdemir
Addresses: I Kara, Osmangazi University, Engineering and Architecture Faculty, Industrial Engineering Department, Eskisehir, Turkey
M.S. Ozdemir, The University of Michigan, College of Engineering, Department of Industrial and Operations Engineering, IOE Building, 1205 Beal Avenue, Ann Arbor, MI 48109-2117 USA.

Title: Development of Timetabler for University Using Heuristic Algorithm
Authors: M. Kim and TaeChoong Chung
Address: Department of Computer Engineering, KyungHee University, YouInSi, KyungKiDo 449-701, Korea.

Title: Repairing University Timetables Using Genetic Algorithms and Simulated Annealing
Authors: N. Mamede and T. Rente
Addresses: N. Mamede, IST/INESC, Technical University of Lisbon, Rua Alves Redol 9, 1000 Lisboa, Portugal
T. Rente, Rua Joaquim Antonio de Aguiar, 66 - 4, 1070 Lisboa, Portugal.

Title: Towards a language for the specification of timetabling problems
Authors: J. M. Mata, A.L. Senna, M.A. Andrade
Address: Departmento de Ciência da Computação, Universidade Federal de Minas Gerais, 31270-010 Belo Horizonte, Minas Gerais, Brasil.

Title: What Is Timetabling Really About?
Author: J. Melis
Address: Infosilem Inc, Academic Management Systems, 100 rue Morin, P.O. Box 3157, Ste. Adèle, Québec JOR 1LO, Canada.

Title: Scheduling a Major College Basketball Conference
Authors: G. L. Nemhauser and M. A. Trick
Address: G. L. Nemhauser, Industrial And Systems Engineering, Georgia Institute of Technology, Atlanta, GA 30332-0205, USA.
M. A. Trick, Graduate Schol of Industrial Administration, Carnegie Mellon University, Pittsburgh, PA 15213-3890, USA.

Title: High School Timetabling By Constraint Programming
Authors: G. Pesant, R. Séguin and P. Soriano
Addresses: Centre de Recherche sur les Transports, Université de Montréal, Canada.

Title: Automated Timetable Generation through Distributed Negotiation
Authors: V. Ram and P. Warren
Addresses: Department of Computer Science and Information Systems, University of Natal, South Africa.

Title: CELCAT: A Practical Solution To Scheduling Problems
Author: S. Rogalla
Address: Corbett Engineering Ltd, 1 Ashfield Road, Kenilworth, Warwickshire,CV8 2BE, UK.

Title: Design and Implementation of a Timetable System using Genetic Algorithm
Authors: A Mendes dos Santos, E. Marques and L. Satoru Ochi
Address: Department de Ciência da Computação, Universidade Federal Fluminense, Praca do Valonguinho s/n - Centro - Niterói, Rio de Janeiro, Brasil.

Title: Combining Local Search and Look-Ahead for Scheduling and Timetabling Problems
Authors: A. Schaerf
Address: Dipartimento di Informatica e Sistemistica, Università di Roma "La Sapienza," Via Salaria 113, 00198, Rome, Italy.

Title: Historial Development, Present Situation and Future Perspectives On Sports Timetabling
Author: J.A.M. Schreuder
Address: University of Twente, TW-STOR. Postbus 217, 7500 AE Enschede The Netherlands.

Title: CRAM Room Scheduling System
Author: T. Shaver
Address: Ad Astra Information Systems, 4300 Shawnee Mission Parkway, Shawnee Mission, Kansas, USA.

Title: Micro-Opportunistic Timetabling
Authors: P. Soares and N. Mamede
Address: IST/INESC. Rua Alves Redo 9, 1000 Lisboa, Portugal.

Title: Real World Timetabling: A Pragmatic View
Author: M. Stüber
Address: Stüber Software, Neuwiesenweg 18, D-56566 Neuwied, Germany

Title: Modeling Of The Classroom Assignment Problem
Author: J. Torres-Jiménez
Address: ITESM Campus Morelos, División de Ingeniería y Ciencias, Reforma 182-A Colonia Lomas de Cuernavaca, Cuernavaca, Morelos, Mexico.

Title: CTS: A Complete Timetabling System
Author: G. White
Address: School of Information Technology and Engineering, University of Ottawa, Ottawa K1N 6N5, Canada.

Author Index

Lecture Notes in Computer Science

For information about Vols. 1–1404

please contact your bookseller or Springer-Verlag